The Analysis of Proximity Data

KENDALL'S ADVANCED THEORY OF STATISTICS
and
KENDALL'S LIBRARY OF STATISTICS

Advisory Editorial Board: PJ Green, University of Bristol; RJ Little, University of Michigan; JK Ord, Pennsylvania State University

The development of statistical theory in the past fifty years is faithfully reflected in the history of the late Sir Maurice Kendall's volumes THE ADVANCED THEORY OF STATISTICS. The ADVANCED THEORY began life as a two volume work (Volume 1, 1943; VOLUME 2, 1946) and grew steadily, as a single authored work, until the late fifties. At that point, Alan Stuart became co-author and the ADVANCED THEORY was rewritten in three volumes. When Keith Ord joined in the early eighties, Volume 3 became the largest and plans were developed to expand, yet again, to a four-volume work. Even so, it became evident that there were gaps in the coverage and that it was becoming increasingly difficult to provide timely updates to all volumes, so a new strategy was devised.

In future, the ADVANCED THEORY will be in the form of three core volumes together with a series of related monographs called KENDALL'S LIBRARY OF STATISTICS. The three volumes of ADVANCED THEORY will be:

1   Distribution Theory
2A Classical Inference and Relationships
2B Bayesian Inference (a new companion volume by Anthony O'Hagan)

KENDALL'S LIBRARY OF STATISTICS will encompass the areas previously appearing in the old Volume 3, such as sample surveys, design of experiments, multivariate analysis and time series as well as non-parametrics and log-linear models, previously covered to some extent in Volume 2. In the preface to the first edition of THE ADVANCED THEORY Kendall declared that his aim was 'to develop a systematic treatment of [statistical] theory as it exists at the present time' while ensuring that the work remained 'a book on statistics, not on statistical mathematics'. These aims continue to hold true for KENDALL'S LIBRARY OF STATISTICS and the flexibility of the monograph format will enable the series to maintain comprehensive coverage over the whole of modern statistics.

Published volumes: 1. MULTIVARIATE ANALYSIS Part 1 Distributions, Ordination and Inference, WJ Krzanowski (University of Exeter) and FHC Marriott (University of Oxford)   1994   0 340 59326 1

2. MULTIVARIATE ANALYSIS Part 2 Classification, Covariance Structures and Repeated measurements, WJ Krzanowski (University of Exeter) and FHC Marriott (University of Oxford)   1995   0 340 59325 3

3. MULTILEVEL STATISTICAL MODELS Second Edition, Harvey Goldstein (University of London)   1995   0 340 59529 9

4. THE ANALYSIS OF PROXIMITY DATA, Brian Everitt (Institute of Psychiatry) and Sophia Rabe-Hesketh (Institute of Psychiatry)   1997   0 340 67776 7

THE ANALYSIS OF PROXIMITY DATA

B. S. Everitt and S. Rabe-Hesketh

Department of Biostatistics and Computing, Institute of Psychiatry, UK

A member of the Hodder Headline Group

LONDON • SYDNEY • AUCKLAND

Copublished In North, Central and South America by
John Wiley & Sons, Inc., New York • Toronto

First published in Great Britain 1997 by Arnold,
a member of the Hodder Headline Group
338 Euston Road, London NW1 3BH

Copublished in North, Central and South America by
John Wiley & Sons Inc., 605 Third Avenue, New York, NY 10158-0012

British Library Cataloguing in Publication Data
A catalogue record for this book is available from the British Library

Library of Congress Cataloging-in-Publication Data
A catalog record for this book is available from the Library of Congress

ISBN 0 340 67776 7

ISBN 0 470 19472 7 (Wiley)

Typeset in 10/11pt Times by Alden Bookset, Didcot, Oxon
Printed in Great Britain by St Edmundsbury Press,
Bury St Edmunds, Suffolk and Bookcraft, Bath.

To Simon (SRH), Hywel and Rachel (BSE)

Contents

Preface

Proximity data consist of measures of similarity or dissimilarity between members of a set of stimuli, individuals or objects of interest, and occur in many different disciplines, particularly psychology, sociology and market research. In some instances such data arise from calculations carried out on the usual multivariate data matrix, the elements of which record the values of a number of variables on a number of 'individuals'. In other circumstances, proximity data are collected directly from experiments in which human subjects are asked to make judgements about the similarity or dissimilarity of pairs of stimuli.

Uncovering the pattern or structure in this type of data may be important for a number of reasons, in particular for discovering the dimensions on which similarity judgements are made. In this text a variety of methods, which are helpful in investigating and exploring proximity data, are described and their use illustrated on a range of data sets. Our hope is that the material contained in the book will be a helpful introduction to this area both for research workers who are not primarily statisticians but who collect and wish to analyse proximity data, and for applied statisticians interested in the underlying methodology.

We would like to thank Forrest Young and Jim Ramsey for being extremely helpful (and prompt) when asked about data and software.

B. S. Everitt and S. Rabe-Hesketh

1

Proximity Data

Introduction

1.1 The concept of *distance* arises in many areas of data analysis; it is, for example, at the heart of many problems involving multivariate description, multivariate inference or multivariate classification. In the scattergram shown in Fig. 1.1, for example, the relative 'distances' between points are obviously central to recognizing the three distinct 'clusters' of observations present. That this is so is clear without, at this point, attempting any formal or explicit definition of distance, but instead simply appealing to its everyday usage.

An almost (but not quite) synonymous term for distance is *dissimilarity*, which occurs frequently in accounts of particular multivariate techniques, for example, *cluster analysis* (see Everitt, 1993), and also arises in the context of certain types of investigation in psychology (see next section). Dissimilarity and its complement, *similarity*, have the obvious interpretation of measuring how dissimilar or similar are objects, individuals, stimuli or other entities of interest (these terms will be used interchangeably throughout the text). Both dissimilarities and similarities are often referred to as *proximities*, and it is the analysis of proximity data of various kinds that is the central concern in this book. Since similarities can readily be converted into dissimilarities (see later), most accounts of methods in later chapters will, without loss of generality, be given in terms of dissimilarities.

Quantifying proximity

1.2 The type of data with which this book is primarily concerned, namely proximities, can arise in a number of ways, and in a number of forms. Most common (at least to statisticians), is where the proximities are derived from the usual $n \times p$ multivariate data matrix, $\mathbf{X}$, the elements, x_{jk}, of which give the values of p variables for each of n individuals. There are many ways of converting the matrix $\mathbf{X}$ into an $n \times n$ matrix of inter-individual proximities,

Variable 2

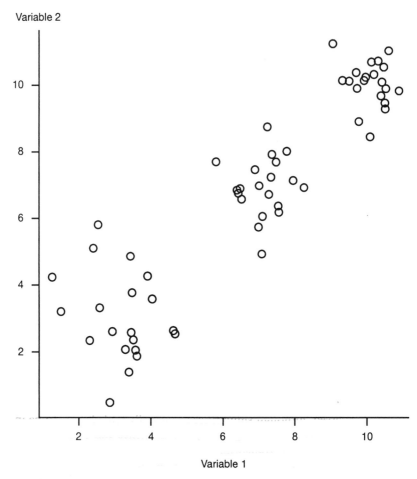

Figure 1.1 Two-dimensional data containing three distinct clusters.

as we shall see in the next chapter. As an example of the process however, Table 1.1 shows a small set of multivariate data and the corresponding matrix containing values of Gower's similarity coefficient (see Chapter 2) for each pair of individuals. The matrix is symmetric and self-similarities are scaled to be unity.

But proximities also frequently arise *directly*, particularly in experiments where people are asked to judge the 'psychological distance' (or 'closeness') of the entities or stimuli of interest. Such experiments are usually designed to uncover rather than impose the dimensions on which human subjects make judgements, so that the attributes on which the stimuli are to be judged are not generally specified. (Subjects can, however, be asked for *specific* kinds of similarity, for example, political or cultural similarity when making judgements about different countries.) As an example, Table 1.2 shows judgements about various brands of cola made by two subjects using a *visual analogue scale* with

Table 1.1 Multivariate data and Gower's similarity coefficient
Consider the following hypothetical data set for five psychiatrically ill patients

	Weight (pounds)	Anxiety level	Depression present?	Hallucinations present?	Age group
Patient 1	120	1	0	0	1
Patient 2	150	2	1	0	2
Patient 3	110	3	1	1	3
Patient 4	145	1	0	1	3
Patient 5	120	1	0	1	1

Anxiety; 1 = mild, 2 = moderate, 3 = severe;
Depression and hallucinations; 0 = no, 1 = yes;
Age group; 1 = young, 2 = middle-aged, 3 = old.
The matrix of Gower's similarity coefficients for these data is **S**, given by

$$\mathbf{S} = \begin{array}{c} \\ 1 \\ 2 \\ 3 \\ 4 \\ 5 \end{array} \begin{pmatrix} \overset{1}{1.000} & \overset{2}{0.062} & \overset{3}{0.150} & \overset{4}{0.344} & \overset{5}{0.750} \\ 0.062 & 1.000 & 0.200 & 0.175 & 0.005 \\ 0.150 & 0.200 & 1.000 & 0.425 & 0.350 \\ 0.344 & 0.175 & 0.425 & 1.000 & 0.475 \\ 0.750 & 0.005 & 0.350 & 0.475 & 1.000 \end{pmatrix}$$

anchor points 'same' (having a score of zero), and 'different' (having a score of one hundred). (Details of the experiment used to obtain the ratings are given in Schiffman *et al.*, 1981.) In this example the resulting rating for a pair of colas is a dissimilarity – low values indicate that the two colas are regarded as alike and vice versa. A similarity measure would have been obtained had the anchor points been reversed, although similarities are usually scaled so as to be in the interval (0, 1) (see Chapter 2). In general, dissimilarities can be converted into similarities by using simple transformations such as

$$\text{dissimilarity} = 1 - \text{similarity} \qquad (1.1)$$

$$\text{dissimilarity} = c - \text{similarity} \qquad (1.2)$$

where c is a suitable constant.

(Human judgements of dissimilarities are of most importance in areas such as psychology and market research and although direct estimates obtained as described above are probably most common, there are a variety of other methods available for collecting such judgements, for example, *preference experiments* and *sorting procedures*, which are described in detail in Schiffman *et al.* 1981, and Rosenberg and Kim, 1975.)

The very nature of the experiment described above imposes symmetry on the matrices of the inter-cola dissimilarity judgements collected. In some situations however, such symmetry is not an integral part of the data. Consider, for example, Table 1.3, which arises from a famous experiment carried out by Rothkopf (1957), in which subjects who did not know Morse code listened to pair of signals (i.e. a sequence of 'dots' and 'dashes'), and were required to state whether the two signals they heard were the same or different. The data in Table 1.3 give the results for the single digit signals and each number in the table

Table 1.2 Dissimilarity data for all pairs of ten colas for two subjects

Subject 1

					Cola number					
	1	2	3	4	5	6	7	8	9	10
1	0									
2	16	0								
3	81	47	0							
4	56	32	71	0						
5	87	68	44	71	0					
6	60	35	21	98	34	0				
7	84	94	98	57	99	99	0			
8	50	87	79	73	19	92	45	0		
9	99	25	53	98	52	17	99	84	0	
10	16	92	90	83	79	44	24	18	98	0

Subject 2

					Cola number					
	1	2	3	4	5	6	7	8	9	10
1	0									
2	20	0								
3	75	35	0							
4	60	31	80	0						
5	80	70	37	70	0					
6	55	40	20	89	30	0				
7	80	90	90	55	87	88	0			
8	45	80	77	75	25	86	40	0		
9	87	35	50	88	60	10	98	83	0	
10	12	90	96	89	75	40	27	14	90	0

is the percentage of a large number of observers who responded 'same' to the row signal followed by the column signal. The point to note about these data is that although the matrix is roughly symmetrical, the symmetry is not exact. The response to signal j following signal i, need not necessarily be the same as that to signal i following j. The proximity matrix in Table 1.3 is *asymmetric*.

Table 1.3 Morse code confusion matrix for the single digits

	1	2	3	4	5	6	7	8	9	0
1 (.- - - -)	84	63	13	8	10	8	19	32	57	55
2 (..- - -)	62	89	54	20	5	14	20	21	16	11
3 (...- -)	18	64	86	31	23	41	16	17	8	10
4 (....-)	5	26	44	89	42	44	32	10	3	3
5 (.....)	14	10	30	69	90	42	24	10	6	5
6 (-....)	15	14	26	24	17	86	69	14	5	14
7 (- -...)	22	29	18	15	12	61	85	70	20	13
8 (- - -..)	42	29	16	16	9	30	60	89	61	26
9 (- - - -.)	57	39	9	12	4	11	42	56	91	78
0 (- - - - -)	50	26	9	11	5	22	17	52	81	94

Table 1.4 Two-way, two-mode data

	G	IT	FR	GB
GC	90	82	88	27
IC	49	10	42	86
TB	88	60	63	99
SS	19	2	4	22
BP	57	55	76	91
SP	51	41	53	55

Countries
G: Germany, IT: Italy, FR: France, GB: Great Britain.

Foods
GC: ground coffee, IC: instant coffee, TB: tea bags,
SS: artificial sweetner, BP: packaged biscuits, SP: package soup.

A further type of proximity data often encountered in practice arises when proximities are available for pairs of stimuli belonging to two different sets. Such data occur as a rectangular matrix with the rows being associated with the stimuli in one set and the columns with the stimuli in the second set. An example is shown in Table 1.4. Here the two sets of 'stimuli' are countries and foods, and the entries in the table give the percentages of households in a country having a particular type of food when surveyed.

A classification of proximity data

1.3 Various classification schemes have been proposed for the type of data of interest in this text, for example, those suggested by Carroll and Arabie (1980) and Jacoby (1991). Here a brief account of the former is given.

Proximity data can be classified in terms of their *ways* and *modes*. The former refers to the number of dimensions of the data array, and the second to the number of different sets of stimuli to which the ways of the data matrix refer. The most common types of proximity data can now be labelled as follows.

(1) *Two-way, one mode.* Two-dimensional, square proximity matrices in which both rows and columns refer to the same stimuli. For examples see the similarity matrix in Table 1.1 and the individual inter-cola dissimilarity matrices in Table 1.2. The analysis of such data is considered in Chapter 3.

(2) *Three-way, two mode.* Three-dimensional proximity matrices in which the rows and columns of each 'slice' refer to the same stimuli and the slices often arise from the judgements of different individuals. This type of data is illustrated in Fig. 1.2. An example is the set of dissimilarity matrices collected in the experiment involving various brands of cola – see Table 1.2. Suitable methods for the analysis of such data are described in Chapter 5.

(3) *Two-way, two-mode.* Two-dimensional proximity matrices in which the rows and columns refer to different sets of stimuli. Such matrices are usually

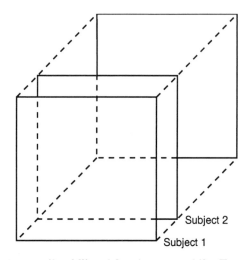

Figure 1.2 Three-way, two-mode data.

rectangular and an example is given in Table 1.4. (An asymmetric proximity matrix can be regarded as a special case of this type of data in which the rows and columns refer to the same stimuli.) Aysmmetric and rectangular proximity matrices are considered in Chapter 6.

Further criteria for classifying proximity data (eg. conditionality) will be described as they arise.

Models for the analysis of proximities

1.4 Models are fitted to proximities in order to clarify, display and possibly explain any structure or pattern not readily apparent in the collection of numerical values. In some areas, particularly psychology, the ultimate goal in the analysis of a set of proximities is more specifically the development of theories for explaining similarity judgements, or, in other words, finding an answer to the question 'what makes things seem alike or seem different?' According to Carroll and Arabie (1980) and Carroll *et al.* (1984), models for the analysis of proximity data can be categorized into one of three major classes, *spatial models, tree models* and *hybrid models.*

Spatial models

1.4.1 Continuous spatial models of the type generally associated with multi-dimensional scaling (see Chapters 3, 5 and 6), usually embed the objects or stimuli of interest in some coordinate space so that a specified measure of distance, for example, *Euclidean* (see Chapter 2), between the points in the space represents the observed proximities. Indeed, a narrow definition of multidimensional scaling (often abbreviated to MDS), is the search for a low-dimensional space, in which points in the space represent the stimuli, one point representing one stimulus, such that the distances between the points in the space $\{d_{ij}\}$, match as well as possible, in some sense, the original dissimilarities $\{\delta_{ij}\}$, or similarities $\{s_{ij}\}$. In a very general sense this simply means that the larger the observed dissimilarity value (or the smaller the similarity value) between two stimuli, the further apart should be the points representing them in the spatial solution. More formally, distances in the derived space are specified to be related to the corresponding dissimilarities in some simple way, for example, linearly, so that the proposed model can be written as follows

$$\delta_{ij} = f(d_{ij}) \qquad\qquad (1.3)$$

$$d_{ij} = h(\mathbf{x}_i, \mathbf{x}_j) \qquad\qquad (1.4)$$

where $\mathbf{x}_i$ and $\mathbf{x}_j$ are q-dimensional vectors containing the coordinate values representing stimuli i and j, f represents the assumed functional relationship between the observed dissimilarities and the derived distances and h represents the chosen distance function. In the majority of applications of MDS, h is taken, often implicitly, to be Euclidean, a point that will be discussed in more detail in both Chapters 2 and 3.

The problem now becomes one of estimating the coordinate values to represent the stimuli. In general this is achieved by optimizing some goodness (badness) of fit index measuring how well the fitted distances match the observed proximities. Details of the procedure will be given in Chapter 3, but to illustrate the type of results given by MDS, Fig. 1.3 shows the results of applying non-metric multidimensional scaling (see Chapter 3) to the Rothkopf Morse code data shown in Table 1.3 using the average of the off-diagonal cells from the original data as the measure of similarity. The possible interpretations of this diagram will be left until Chapter 4.

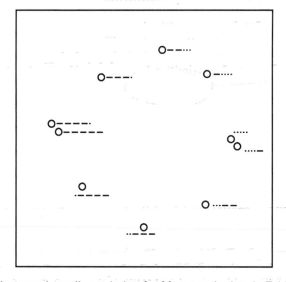

Figure 1.3   Non-metric scaling solution for Morse code data in Table 1.3.

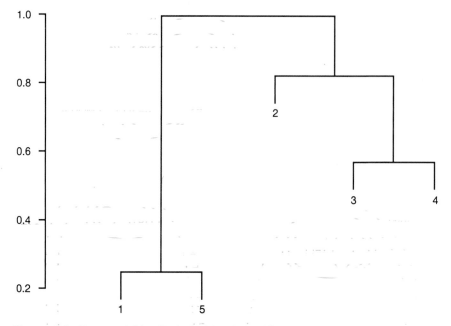

Figure 1.4   Tree model for dissimilarities derived from Gower's similarity coefficient given in Table 1.1.

Non-spatial models

1.4.2   The most common type of non-spatial model for representing the relationships between a set of similarities or dissimilarities is a tree, where each stimulus is represented as a terminal node in a connected graph without cycles (these terms are defined in Chapters 4 and 7). Each pair of nodes is joined by a unique path so that the relationships between the terminal nodes in the graph reflect the observed proximity relations amongst the stimuli. Such a tree structure lends itself to a natural interpretation as a hierarchical clustering scheme (see Johnson, 1967; Gordon, 1987, 1996; Everitt, 1993). Tree models will be considered in detail in Chapter 7, but for interest here, Fig. 1.4 shows a possible tree representation of the dissimilarities produced by taking one minus the similarity values in Table 1.1.

Hybrid models

1.4.3   Hybrid models for proximity data combine aspects of tree structure with a spatial component. Using both a tree model and a spatial model simultaneously may assist in reaching a more complete understanding of any structure underlying the observed proximities. At one level, a 'hybrid model' may consist of nothing more than the combination of the separate results of applying multidimensional scaling and hierarchical cluster analysis to the observed proximity matrix. In investigating the strengths of mental associations among 16 familiar kinds of animals, for example, Shepard (1974) started with the

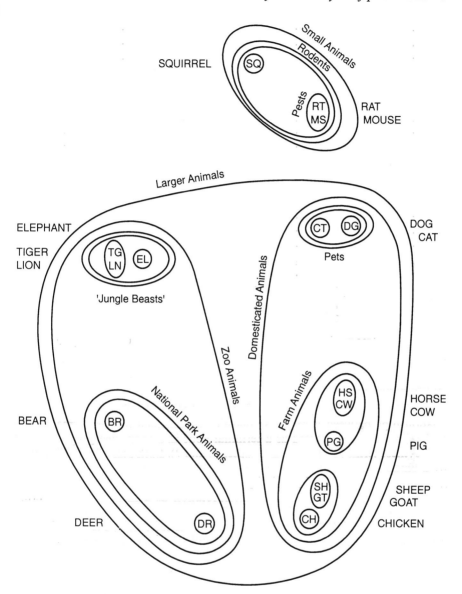

Figure 1.5   Mental associations among 16 kinds of animals; MDS solution plus hierarchical clustering solution (reproduced with permission from Shepard, 1974).

quantitative information from a multidimensional scaling solution and subsequently obtained typal information by performing a hierarchical cluster analysis. Shepard gained substantially increased insight into the data structure after superimposing the typal information on to the derived spatial configuration by enclosing cluster members within closed surfaces – see Fig. 1.5.

Degerman (1970) proposed the first formal hybrid model combining elements of continuous dimensional structure and of discrete class-like structure. The model involves the rotation of a high-dimensional MDS solution so as to find subspaces in which there is class-like rather than continuous variation. Other such hybrid models have been proposed by Carroll (1976), Carroll and Pruzansky (1980) and De Sarbo *et al.* (1990).

Distances and classical multivariate analysis

1.5   The classical approaches to the analysis of multivariate data as exemplified in techniques such as principal components analysis and factor analysis, usually stress the study of the covariance structure amongst the variables as the prime objective, with the objects or individuals on which the variables are measured being regarded as merely a replication factor for obtaining the covariance measures. But such methods may also be viewed as attempts to achieve a low dimensional representation of the data.

Originally, each individual in the data set can be thought of as being represented as a point in a space having as many dimensions as there are variables. The objective of, for example, principal components analysis, is then that of finding coordinates in a lower dimensional representation space so that distances among the individuals in the original space are approximated well by the corresponding distances in the low dimensional space. Such representations are often referred to as *ordinations* of the data.

Formulated as described above, the aim of a technique such as principal components analysis is seen to be similar to that of multidimensional scaling, as outlined in Section 1.4.1. Although the variables may now appear to play a secondary role, since the distances to be approximated are merely derived from them, their influence in the analysis is crucial, since their covariance structure has implications for the appropriate distance measure to consider, a point that will be taken up in detail in Appendix A where the distance properties of classical ordination techniques are considered more fully.

Summary

1.6   The two goals of the analysis of proximities are the construction of models for describing and displaying the implied relationships between the objects, individuals or stimuli of interest, and the development of theories of explaining these relationships. Suitable models on which to base such analyses can be roughly divided into three classes, spatial models, tree models and hybrid models. Details of many of these models and examples of their application will be found in later chapters of the text. All such models are based on the concepts of dissimilarity and distance, concepts which will be discussed in detail in Chapter 2.

2

Measures of Similarity, Dissimilarity and Distance

Introduction

2.1 In Chapter 1 the terms distance, dissimilarity and similarity were introduced in an intuitive and informal manner. In this chapter these concepts will be discussed more formally and the differences between them clarified. Since similarity coefficients can be transformed relatively simply into dissimilarity coefficients (see Chapter 1), much of the discussion in this chapter will involve only dissimilarities and distances, and in this text the distinction between the two terms will involve the fundamental notion of a *metric*, which can be defined in the following way.

Consider an $n \times n$ matrix of dissimilarities, Δ, with elements, δ_{ij}, where $\delta_{ii} = 0$ for all i, then Δ is said to be metric, if the metric (triangular) inequality

$$\delta_{ij} + \delta_{ik} \geqslant \delta_{jk} \tag{2.1}$$

holds for all triplets (i, j, k).

Gower and Legendre (1986) demonstrated that some simple but important properties follow from this definition. Consideration of the triplet (i, j, j) shows that $\delta_{ij} \geqslant 0$ for all pairs (i, j). Consideration of (i, j, i) and (j, i, j) shows that $\delta_{ij} \geqslant \delta_{ji}$ and $\delta_{ji} \geqslant \delta_{ij}$. Hence, all metric dissimilarity matrices are symmetric with non-negative elements. Suppose $\delta_{ij} = 0$ then considering the triplets (i, k, j) and (j, k, i) yields that $\delta_{ik} = \delta_{jk}$ for all k. This is a basic property of metrics that can be strengthened to show that if two points are close together (δ_{ij} close to zero), then any third point, k, will have a similar relation to both of them (i.e. δ_{ik} and δ_{jk} will be of similar size).

In this text, dissimilarity measures which are metric will be explicitly referred to as distance measures. (A matrix of distance measures will generally be denoted by $\mathbf{D}$, and its elements by d_{ij}.) Probably the most important, and certainly the most familiar distance measure is Euclidean, which in its basic form is defined as follows

$$d_{ij} = \left(\sum_{k=1}^{p} (x_{ik} - x_{jk})^2 \right)^{1/2} \tag{2.2}$$

where x_{ik} and x_{jk} are, respectively, the kth coordinates of points i and j in a p-dimensional space.

A Euclidean metric solution is the one most commonly sought by models attempting to represent and display proximity data, but many other distance measures are available as will be seen in the next section, and not all metrics are necessarily Euclidean – where a distance matrix, **D**, is said to be Euclidean if the n points can be embedded in a Euclidean space such that the Euclidean distance between points i and j is d_{ij}. This point is illustrated by Gower and Legendre (1986) with the following distance matrix

$$
\mathbf{D} = \begin{array}{c} \\ 1 \\ 2 \\ 3 \\ 4 \end{array} \begin{array}{c} \begin{array}{cccc} 1 & 2 & 3 & 4 \end{array} \\ \left(\begin{array}{cccc} 0 & & & \\ 2 & 0 & & \\ 2 & 2 & 0 & \\ 1.1 & 1.1 & 1.1 & 0 \end{array} \right) \end{array} \tag{2.3}
$$

It is straightforward to show that **D** is metric, simply by verifying the triangle inequality for all triplets. The matrix arises from a situation in which points, 1, 2 and 3 form an equilateral triangle of side 2, with point 4 being equidistant (1.1 units) from each of 1, 2 and 3. Now, if this configuration is to be Euclidean then the smallest distance that point 4 can be from the other points is when it is coplanar with them and at their centroid. But this corresponds to a minimal distance of $(2\sqrt{3})/3 = 1.15$, which is greater than 1.1. Thus **D** is metric but not Euclidean. This example is illustrated in Fig. 2.1.

The almost universal aim of seeking a Euclidean solution when applying, in particular, spatial models for proximity data, probably reflects our own experience with such a metric living as we do (locally at least) in a Euclidean environment. Such an aim is, however, not always necessary, and may, in certain cases be misplaced, as will be seen in the next chapter. It is certainly

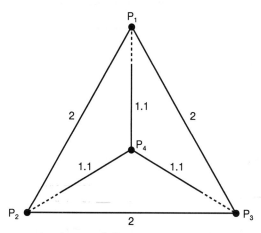

Figure 2.1 An example of a set of distances that satisfy the metric inequality but which have no Euclidean representation.

unlikely that dissimilarities obtained directly on the basis of subjective judgements about stimuli (for example, those shown in Table 1.2) will be Euclidean; indeed it is even unlikely that they will be metric. But what about distance and dissimilarity measures obtained indirectly from multivariate data? It is clearly of importance to consider for which of these derived measures a Euclidean representation is possible, and this question is taken up in the next section. (For the moment it will be assumed that it is measures of dissimilarity between unique individuals or stimuli that are of interest; measures for the situation where the 'entities' under consideration are well-defined populations from which individuals may be sampled will be considered in Section 2.4.)

Derived similarity and dissimilarity measures

2.2 To begin, dissimilarity measures for data where the variables are quantitative (continuous, possibly discrete but not binary) will be discussed; following this, the many similarity coefficients suggested for binary data will be reviewed. Lastly in this section, possible dissimilarity measures for mixed mode data, i.e. data containing both quantitative and binary variables, will be considered.

Dissimilarity measures derived from quantitative data

2.2.1 Many methods have been proposed for proceeding from a set of multivariate observations describing a sample of individuals or stimuli of interest, to a dissimilarity matrix for the individuals; the most common of these are listed in Table 2.1.

Euclidean distance (D1) was mentioned previously in Section 2.1, but without any comment about the problem of standardization. The latter is, however, of great practical concern since calculation of a distance depends on the scales and

Table 2.1 Dissimilarity measures for quantitative data

Measure	Formula
D1: Euclidean distance Unstandardized: $w_k = 1$ Standardized by SD: $w_k = 1/s_k^2$	$d_{ij} = \left(\sum_{k=1}^{p} w_k (x_{ik} - x_{jk})^2 \right)^{1/2}$
D2: City block	$d_{ij} = \sum_{k=1}^{p} |x_{ik} - x_{jk}|$
D3: Minkowski	$d_{ij} = \left\{ \sum_{k=1}^{p} |x_{ik} - x_{jk}|^r \right\}^{1/r}$
D4: Canberra (Lance and Williams, 1966)	$d_{ij} = \sum_{k=1}^{p} \frac{|x_{ik} - x_{jk}|}{|x_{ik} + x_{jk}|}$
D5: Bray–Curtis (Bray and Curtis, 1957)	$d_{ij} = \frac{\sum_{k=1}^{p} |x_{ik} - x_{jk}|}{\sum_{k=1}^{p} (x_{ik} + x_{jk})}$

inclination of the axes. In the case of unique entities, normalizing each variable by dividing by its standard deviation over the n individuals is often recommended, although some authors (for example, Carmichael *et al.*, 1968), suggest scaling by the range. Such scaling may be perfectly sensible (and necessary) in some circumstances but not in all; Fleiss and Zubin (1969), for example, highlight its deficiencies in the context of cluster analysis, where any scaling should, ideally, be performed on individual clusters, rather than the complete set of data.

(Scaling is also an issue when the rows of the data matrix X are considered as 'profiles' characterizing each individual, and correlations between pairs of individuals are used to quantify their similarity; consequently rows are standardized instead of columns. This practice has been criticized by many authors including Jardine and Sibson, 1971, and Everitt, 1993, although it can be partially justified when the question of the 'shape' of profiles is of more interest than those questions concerning level and scale. An example of using the correlation coefficient in this way when the profiles are time series is given in Chapter 3.)

Problems also arise when using (2.2) to calculate distances when the variables are correlated, and Sokal (1961) has pointed out that such calculation does not lead to Euclidean distances. A measure of distance to be described in Section 2.4, *Mahalanobis D^2*, does allow for correlations between variables.

The city block measure (D2) is also known as the *taxicab* (Krause, 1975), *rectilinear* (Brandeau and Chiu, 1988) or *Manhattan* (Larson and Sadiq, 1983) distance because it describes distances travelled in such a street configuration. More formally it is often referred to as the l_1 norm.

Both Euclidean distance and city block distance are examples of the more general Minkowski measure (D3) (Euclidean corresponds to $r = 2$, city block to $r = 1$). All such measures are metric.

Of the family of Minkowski metrics, only in the Euclidean case are distances invariant under orthogonal rotations of the coordinate axes. All such metrics, however, possess invariance under translations and under central reflections of the coordinate axes (see Busemann, 1950, 1955, and de Leeuw and Heiser, 1982). The implications of such invariances when fitting models to proximity data will be taken up later, but it is of interest to note here that Arabie (1991), following comments by Attneave (1950), uses the lack of invariance to orthogonal rotations of the city block metric, as one of the arguments in favour of its use over the Euclidean measure for representing psychological distances (see Chapter 3).

The metric (or non-metric) characteristics of the remaining measures in Table 2.1 are derived in Gower and Legendre (1986). Measure D4, for example, can be shown to be metric for positive values of the variables but not metric when these values can be negative. Measure D5 is non-metric for both positive and negative values.

A further question about derived distance measures, which is of interest, particularly in the context of multidimensional scaling, is whether or not they are Euclidean. Gower and Legendre (1986) show that, apart from Euclidean distance itself (standardized as well as unstandardized), only measure D4 has this property, and then only when the variable values are positive. The implications of whether or not a measure is Euclidean when modelling proximity data will be discussed further in later chapters.

Table 2.2 Similarity coefficients for binary data

Measure	Formula
S1: Matching coefficient	$s_{ij} = \dfrac{a+d}{a+b+c+d}$
S2: Jaccard coefficient (Jaccard, 1908)	$s_{ij} = \dfrac{a}{a+b+c}$
S3: Rogers and Tanimoto (1960)	$s_{ij} = \dfrac{a+d}{a+2(b+c)+d}$
S4: Gower and Legendre (1986)	$s_{ij} = \dfrac{a-(b+c)+d}{a+b+c+d}$
S5: Sokal and Sneath (1963)	$s_{ij} = \dfrac{a}{a+2(b+c)}$
S6: Gower and Legendre (1986)	$s_{ij} = \dfrac{a+d}{a+1/2(b+c)+d}$
S7: Gower and Legendre (1986)	$s_{ij} = \dfrac{ad-bc}{ad+bc}$

Similarity measures for binary data

2.2.2 When all the variables describing the individuals are binary (a common occurrence in many areas such as psychiatry, biology etc.), it is traditional to consider similarity rather than dissimilarity measures, and this is the course followed in this section. (Transformation of the derived similarity measures into dissimilarities can be accomplished as suggested in Chapter 1.)

A list of many of the similarity measures that have been proposed for binary data is given in Table 2.2. The measure of similarity for two individuals, i and j is based on the following table of counts of matches and mismatches in the two sets of variable values.

		Individual i		
		1	0	Total
Individual j	1	a	b	$a+b$
	0	c	d	$c+d$
Total		$a+c$	$b+d$	$p=a+b+c+d$

For some binary variables such as 'sex' there is, of course, no preference which outcome should be coded as 0 and which as 1. In other cases the zero category corresponds to the genuine absence of some property and the plethora of similarity measures for binary data arises largely because of uncertainty over how to deal with the co-absences of a particular variable. The question is largely dependent on the answer to the question of whether such co-absences contain useful information about the similarity of two objects. In many applications, attributing a large degree of similarity to a pair of individuals simply because they both lack a large number of attributes might be considered inappropriate. If, for example, the presence or absence of a relatively rare attribute, such as blood type AB negative, is of interest, two people with blood that is AB negative clearly have something in common, but it is not so clear that this can be said of two people who do not have the condition. In such cases a measure of similarity that ignores the co-absence count, d, for example, Jaccard's coefficient (S2), might be deemed more suitable. When co-absences are deemed relevant, the simple matching coefficient (S1) is usually employed (the square root of one minus the simple matching coefficient is equivalent to Euclidean distance for binary data). Measures with intermediate weightings of a and d (S3 and S6, for example) have been criticized on the basis that they involve unjustified scaling.

Gower and Legendre (1986) prove the following theorem concerning the Euclidean properties of similarity coefficients. If S, a similarity matrix with elements $0 \leqslant s_{ij} \leqslant 1$ and $s_{ii} = 1$, is positive semi-definite, then the dissimilarity matrix with elements, $d_{ij} = (1 - s_{ij})^{1/2}$ is Euclidean. The theorem gives only a sufficient condition for $(1 - s_{ij})^{1/2}$ to be Euclidean, but, fortunately, for many similarity coefficients, S may be shown to be positive semi-definite and hence the question of necessity becomes irrelevant.

Gower and Legendre (1986), demonstrate that for the majority of similarity coefficients given in Table 2.2 the corresponding similarity matrix is positive semi-definite. Only for coefficients S6 and S7 does the condition not hold.

Proximity measures for data containing both quantitative and binary variables

2.2.3   At least three proximity measures have been described that handle mixtures of different kinds of variables (Estabrook and Rodgers, 1966; Gower, 1971; Legendre and Chodorowski, 1977). Other mixed coefficients might be constructed relatively simply by combining coefficients from Tables 2.1 and 2.2, either with, or without, differential weighting. Here, only Gower's coefficient will be considered in detail.

Gower (1971) introduced a general similarity measure of the form

$$s_{ij} = \frac{\sum_{k=1}^{p} w_{ijk} s_{ijk}}{\sum_{k=1}^{p} w_{ijk}} \tag{2.4}$$

where s_{ijk} is the similarity between individuals i and j based only on the kth variable (for binary variables and qualitative variables with more than two categories, $s_{ijk} = 1$ if the kth variable is the same for individuals i and j and zero otherwise), and w_{ijk} is unity if individuals i and j can legitimately be compared on the kth variable and zero if they cannot. If, for example, the kth variable value is missing for either or both of i and j, the value of w_{ijk} is set to zero. In addition if, when considering binary variables, it is thought appropriate to ignore co-absences on a variable, the corresponding 'w' value is taken to be zero. For quantitative variables, Gower suggests the following

$$s_{ijk} = 1 - |x_{ik} - x_{jk}|/R_k \tag{2.5}$$

where R_k is the range of the observations for variable k.

Gower (1971) shows that when there are no missing values the similarity coefficient above is positive semi-definite; consequently, points can be found in Euclidean space with distances $(1 - s_{ij})^{1/2}$. Missing values may cause the similarity matrix to lose its positive semi-definite property, as can be seen from the following example in which there are three individuals and four variables.

Individual	Variable			
	1	2	3	4
1	−	−	+	+
2	+	+	+	*
3	+	+	+	+

Here, * denotes a missing value and $+/-$ may represent either the presence/absence of dichotomous variates or alternative values of qualitative variates. In either case the derived Gower's similarity matrix is

$$S = \begin{pmatrix} 1 & \frac{1}{3} & \frac{1}{2} \\ \frac{1}{3} & 1 & 1 \\ \frac{1}{2} & 1 & 1 \end{pmatrix} \tag{2.6}$$

The determinant of S is $-1/36$ and so S is not positive semi-definite.

Choice of measure

2.3 Gower and Legendre (1986) make the important point that the nature of the data should strongly influence the choice of a similarity or dissimilarity measure. Under certain circumstances, for example, quantitative data may best be regarded as binary as when dichotomizing 'noisy' quantitative variables (see Legendre and Legendre, 1983), or when the relevant information for the purpose an investigator has in mind depends on a known threshold value. An example, given by Gower and Legendre (1986), involves classifying river areas according to their suitability for growing edible fish as judged by threshold levels of pesticides and heavy metals; in such a case it makes good sense to code the data in binary form, according to whether measurements are above or below some tolerated toxicity level, despite the fact that the measurements themselves are quantitative.

When choosing a similarity measure for binary data, the choice is generally between the simple matching coefficient and Jaccard's coefficient (despite the large number of other measures that have been suggested for such data). As mentioned in the previous section the decision as to which of these two measures to use will be based on how the investigator wishes to handle co-absences. For data that involve a mixture of quantitative and binary variables (and perhaps categorical variables with more than two categories), Gower's similarity coefficient is often very useful.

It might be thought that the method of analysis to be applied would have some implications for the choice of similarity or dissimilarity measure.

Multidimensional scaling methods that have a geometric rationale, for example, might appear to require measures that are metric and possibly even Euclidean. But even with methods, like classical scaling, (which as we shall see in Chapter 3, appear to require a Euclidean measure) a modest departure may be inconsequential, a point which will be taken up again in later chapters.

Gower and Legendre (1986) discuss in some detail choosing a similarity or dissimilarity measure and give a decision-making table that may often be helpful in the process. Overall, however, they conclude that it is not possible to give a definitive answer to what measure is best to use in all circumstances.

Inter-group distance and dissimilarity measures

2.4   In the previous sections of this chapter it has been measures of dissimilarity and distance between individuals that have been of concern. There are, however, many situations that arise in practice, in which these concepts have to be considered for groups of individuals or even for populations described by some particular probability density function. As with measures for individuals it is helpful to consider inter-group measures first for quantitative variables and then for binary variables.

Inter-group measures for quantitative variables

2.4.1   It was in 1936 that Mahalanobis introduced his now famous *generalized distance measure* for assessing the distance between two p-variate normal populations with the same covariance matrix, Σ. The measure is defined as

$$D^2 = (\mu_1 - \mu_2)' \Sigma^{-1} (\mu_1 - \mu_2) \tag{2.7}$$

where μ_1 and μ_2 are the p-variate mean vectors of the two populations.

(Rao, 1949, considered the more general problem of defining a distance between any two members of an arbitrary family of probability distributions, a topic considered in detail by Krzanowski and Marriot, 1995.)

Mahalanobis distance can be applied to two groups of individuals by assuming that these represent independent samples from two normal populations, and replacing μ_1, μ_2 and Σ by their sample estimates, i.e. the sample mean vectors $\bar{x}_1$ and $\bar{x}_2$, and the estimated common covariance matrix, S, given by

$$S = \frac{(n_1 - 1)S_1 + (n_2 - 1)S_2}{n_1 + n_2 - 2} \tag{2.8}$$

where S_i, $i = 1, 2$, represent the sample covariance matrices of each group, and n_i, $i = 1, 2$, the sample sizes. Mahalanobis distance underlies canonical variate analysis (see Appendix A).

The user of Mahalanobis distance makes the assumption that the two normal populations from which independent samples are available have the same covariance matrix. For samples with heterogeneous covariance matrices, a number of alternatives to Mahalanobis distance have been suggested. Three such alternatives were investigated by Chadda and Marcus (1968), who concluded that a measure proposed by Anderson and Bahadur (1962) has some advantages; this measure is defined as

$$\Delta = \max_{t} \frac{2\mathbf{b}'_t\mathbf{d}}{(\mathbf{b}'_t\mathbf{S}_1\mathbf{b}_t)^{1/2} + (\mathbf{b}'_t\mathbf{S}_2\mathbf{b}_t)^{1/2}} \tag{2.9}$$

where $\mathbf{d} = \bar{\mathbf{x}}_1 - \bar{\mathbf{x}}_2$ and $\mathbf{b}_t = (t\mathbf{S}_1 + (1-t)\mathbf{S}_2)^{-1}\mathbf{d}$.

A further possibility for measuring the distance between samples from two normal populations with different covariance matrices is the normal information radius (NIR) suggested by Jardine and Sibson (1971), and given by

$$\text{NIR} = \frac{1}{2}\log_2\left\{\frac{|\frac{1}{2}(\mathbf{S}_1 + \mathbf{S}_2)| + \frac{1}{4}(\bar{\mathbf{x}}_1 - \bar{\mathbf{x}}_2)'(\bar{\mathbf{x}}_1 - \bar{\mathbf{x}}_2)}{|\mathbf{S}_1|^{1/2}|\mathbf{S}_2|^{1/2}}\right\} \tag{2.10}$$

When $\mathbf{S}_1 = \mathbf{S}_2 = \mathbf{S}$ this reduces to

$$\text{NIR} = \frac{1}{2}\log_2\left\{1 + \frac{1}{4}(\bar{\mathbf{x}}_1 - \bar{\mathbf{x}}_2)'\mathbf{S}^{-1}(\bar{\mathbf{x}}_1 - \bar{\mathbf{x}}_2)\right\} \tag{2.11}$$

$$= \frac{1}{2}\log_2\left\{1 + \frac{3}{4}D^2\right\} \tag{2.12}$$

where D^2 is Mahalanobis distance; NIR may therefore be regarded as providing a generalization of D^2 appropriate to the unequal covariance case.

A variety of other inter-group distance measures are used in hierarchical cluster analysis; details are given in Everitt (1993).

Inter-group distance measures for qualitative variables

2.4.2 The problem of measuring the distance between populations using categorical variables has been considered by a number of authors. Balakrishnan and Sanghvi (1968), for example, suggest an index of the form

$$G^2 = \sum_{j=1}^{r}\sum_{k=1}^{s_j+1}\frac{(p_{1jk} - p_{2jk})^2}{p_{jk}} \tag{2.13}$$

where p_{1jk} and p_{2jk} are the proportions in the kth category of the jth variable in the two populations, $p_{jk} = \frac{1}{2}(p_{1jk} + p_{2jk})$, $s_j + 1$ is the number of classes for the jth variable, and r is the number of variables.

Another approach to defining a distance measure based on discrete variables was adopted by Kurczynski (1969) and involved the extension of the Mahalanobis generalized distance, with discrete variables replacing continuous variables. In its most generalized form, the measure for discrete variables is given by

$$D^2 = (\mathbf{p}_1 - \mathbf{p}_2)'\mathbf{S}^{-1}(\mathbf{p}_1 - \mathbf{p}_2) \tag{2.14}$$

where $\mathbf{p}'_i = [p_{i11}, p_{i12}, \ldots, p_{i1s_1}, p_{i21}, p_{i22}, \ldots, p_{i2s_2}, \ldots, p_{ir1}, p_{ir2}, \ldots, p_{irs_r}]$ contains the sample proportions in the ith population, and $\mathbf{S}$ is the $M \times M$ common sample covariance matrix, where $M = \sum_{j=1}^{r} s_j$. Various alternative forms of this distance measure may be derived, depending on how the elements of **S** are calculated. Kurczynski (1970), for example, shows that if each variable has a multinominal distribution, and the variables are independent of one another, then the generalized distance measure in (2.14) is equal to the measure defined in (2.13).

Inter-group distance measures for categorical data find important applications where the variables are gene frequencies; examples are given in Kurczynski (1969).

Summary

2.5 Many different inter-individual dissimilarity measures can be derived from a matrix of variable values for the individuals. Those that satisfy the metric inequality are generally referred to as distance measures. In the case of quantitative variables, Euclidean distance holds a special place but many other distance measures have been used in particular applications. When the variables describing each individual are binary, again a host of similarity measures are available. Choosing between them usually involves deciding how to deal with co-absences. Some of the derived dissimilarity and similarity measures have the property that they may be represented in a Euclidean space, a property which can have implications for multidimensional scaling procedures, as we shall see in the next chapter.

In some applications (particularly those involving clustering), it is distance between populations, or samples of individuals from particular populations, which is of interest. The generalized distance of Mahalanobis has been used for many years to assess the distance between samples from multivariate normal distributions, and analogues for categorical data have been suggested.

3

Spatial Representation of Proximity Data: Metric and Non-metric Multidimensional Scaling

Introduction

3.1 A spatial representation of a proximity matrix consists of a set of q-dimensional coordinates representing each individual or stimulus, chosen so that the distances (however defined) between the points in the q-dimensional space, match closely in some sense the observed dissimilarities or similarities. In general, the distance function assumed will be Euclidean but this is not essential, and in some cases not desirable (see later). Finding both the 'best' fitting set of coordinates and the appropriate value of q is the general aim of multidimensional scaling techniques. The variety of methods that have been suggested differ largely in how agreement between fitted distances and observed proximities is measured, i.e. in their goodness-of-fit (badness-of-fit) criteria. Three classes of scaling technique will be considered in this chapter, classical scaling, least squares scaling (both of which are examples of metric scaling) and non-metric scaling.

Classical multidimensional scaling

3.2 To introduce classical scaling, consider first the small multivariate data matrix, $\mathbf{X}$, given by

$$\mathbf{X} = \begin{pmatrix} 3 & 4 & 4 & 6 & 1 \\ 5 & 1 & 1 & 7 & 3 \\ 6 & 2 & 0 & 2 & 6 \\ 1 & 1 & 1 & 0 & 3 \\ 4 & 7 & 3 & 6 & 2 \\ 2 & 2 & 5 & 1 & 0 \\ 0 & 4 & 1 & 1 & 1 \\ 0 & 6 & 4 & 3 & 5 \\ 7 & 6 & 5 & 1 & 4 \\ 2 & 1 & 4 & 3 & 1 \end{pmatrix}$$

The Euclidean distances between the rows of $\mathbf{X}$ can be calculated by simply applying (2.2) to the five variable values; this leads to the following distance matrix

$$
\mathbf{D} = \begin{pmatrix}
0.00 \\
5.20 & 0.00 \\
8.37 & 6.08 & 0.00 \\
7.87 & 8.06 & 6.32 & 0.00 \\
3.46 & 6.56 & 8.37 & 9.27 & 0.00 \\
5.66 & 8.42 & 8.83 & 5.29 & 7.87 & 0.00 \\
6.56 & 8.60 & 8.19 & 3.87 & 7.42 & 5.00 & 0.00 \\
6.16 & 8.89 & 8.37 & 6.93 & 6.00 & 7.07 & 5.74 & 0.00 \\
7.42 & 9.05 & 6.86 & 8.89 & 6.56 & 7.55 & 8.83 & 7.42 & 0.00 \\
4.36 & 6.16 & 7.68 & 4.80 & 7.14 & 2.64 & 5.10 & 6.71 & 8.00 & 0.00
\end{pmatrix}
$$

$$(3.1)$$

Now suppose that it is $\mathbf{D}$ that is given and the corresponding $\mathbf{X}$ that is required. How can a set of coordinate values be derived from $\mathbf{D}$, such that their Euclidean distances are the elements of $\mathbf{D}$? The answer to this question is provided by classical scaling, which has its origin in the work of Young and Householder (1938).

To begin we must note that there is no unique set of coordinate values which will give rise to the distances in $\mathbf{D}$, since these distances are unchanged by shifting the whole configuration of points from one place to another, or by rotation or reflection of the configuration. In other words, we cannot uniquely determine either the location or the orientation of the configuration. The location problem is usually overcome by placing the mean vector of the configuration at the origin. The orientation problem means that any configuration derived can be subjected to an arbitrary orthogonal transformation. Such transformations are often used to facilitate the interpretation of solutions as will be seen in later examples. The initial step in describing classical scaling is the introduction of the $n \times n$ inner products matrix, $\mathbf{B}$, given by

$$\mathbf{B} = \mathbf{XX}' \qquad (3.2)$$

The elements of $\mathbf{B}$ are given by

$$b_{ij} = \sum_{k=1}^{p} x_{ik}x_{jk} \qquad (3.3)$$

where x_{ik} is the element of $\mathbf{X}$ in the ith row and kth column, p is the number of columns of $\mathbf{X}$ and n the number of rows.

It is easy to see that the squared Euclidean distances between the rows of $\mathbf{X}$ can be written in terms of the b_{ij} as

$$d_{ij}^2 = b_{ii} + b_{jj} - 2b_{ij} \qquad (3.4)$$

Now suppose that only the Euclidean distances, d_{ij}, were known, but from these we could, by some suitable procedure, find the corresponding elements of **B**; the required coordinate values could be derived by factoring **B** in the form shown in (3.2). To obtain the b_{ij} in terms of the d_{ij} involves inverting equation (3.4). No unique solution exists unless a location constraint is introduced; as noted above this is usually to set the centre of gravity of the points, $\bar{x}$, at the origin. Consequently, $\sum_{i=1}^{n} x_{ik} = 0$ for all k. These constraints and equation (3.3) imply that the sum of the terms in any row of **B** must be zero, and so summing (3.4) over i, over j and finally over both i and j, leads to the following series of equations

$$\sum_{i=1}^{n} d_{ij}^2 = T + nb_{jj}$$

$$\sum_{j=1}^{n} d_{ij}^2 = nb_{ii} + T \tag{3.5}$$

$$\sum_{i=1}^{n}\sum_{j=1}^{n} d_{ij}^2 = 2nT$$

where $T = \sum_{i=1}^{n} b_{ii}$ is the trace of the matrix, **B**. From (3.4) and (3.5), the elements of **B** can be found in terms of squared Euclidean distances as

$$b_{ij} = -\tfrac{1}{2}[d_{ij}^2 - d_{i.}^2 - d_{.j}^2 + d_{..}^2] \tag{3.6}$$

where

$$d_{i.}^2 = \frac{1}{n}\sum_{j=1}^{n} d_{ij}^2, \quad d_{.j}^2 = \frac{1}{n}\sum_{i=1}^{n} d_{ij}^2, \quad d_{..}^2 = \frac{1}{n^2}\sum_{i=1}^{n}\sum_{j=1}^{n} d_{ij}^2$$

Having now derived the elements of **B** in terms of Euclidean distances, it remains to factor the matrix into the form shown in (3.2) to find the required coordinate values. This can be achieved by writing **B** in terms of its spectral decomposition as

$$\mathbf{B} = \mathbf{V\Lambda V}' \tag{3.7}$$

where $\mathbf{\Lambda} = \text{diag}[\lambda_1, \lambda_2, \dots, \lambda_n]$ is the diagonal matrix of eigenvalues of **B**, and $\mathbf{V} = [\mathbf{v}_1, \mathbf{v}_2, \dots, \mathbf{v}_n]$, the corresponding matrix of eigenvectors, normalized so that the sum of squares of their elements is unity, i.e. $\mathbf{v}_i'\mathbf{v}_i = 1$. The eigenvalues are assumed labelled such that $\lambda_1 \geqslant \lambda_2 \geqslant \cdots \geqslant \lambda_n$.

When **D** arises from an $n \times p$ matrix of full rank, then the rank of **B** is p, and so the last $n - p$ of its eigenvalues will be zero. Thus **B** can be written in the form

$$\mathbf{B} = \mathbf{V}_1\mathbf{\Lambda}_1\mathbf{V}_1' \tag{3.8}$$

where

$$\mathbf{\Lambda}_1 = \text{diag}[\lambda_1, \lambda_2, \dots, \lambda_p], \quad \mathbf{V}_1 = [\mathbf{v}_1, \mathbf{v}_2, \dots, \mathbf{v}_p]$$

It follows that the required matrix of coordinate values is

$$\mathbf{X} = \mathbf{V}_1\mathbf{\Lambda}_1^{1/2} \tag{3.9}$$

where $\mathbf{\Lambda}_1^{1/2} = \text{diag}[\lambda_1^{1/2}, \lambda_2^{1/2}, \dots, \lambda_p^{1/2}]$

Example 3.1

Applying the classical scaling procedure described above to the matrix of Euclidean distances, **D**, given at the beginning of this section, leads to the following derived coordinate values

$$\begin{pmatrix}
-1.60 & 2.38 & -3.23 & 0.37 & 0.12 \\
-2.82 & -2.31 & -3.95 & 0.34 & 0.33 \\
-1.69 & 5.14 & 1.29 & 0.65 & -0.05 \\
3.95 & -2.43 & 0.38 & 0.69 & 0.03 \\
-3.60 & 2.75 & -0.26 & 1.08 & -1.26 \\
2.95 & 1.35 & -0.19 & -2.82 & 0.12 \\
3.47 & 0.76 & 0.30 & 1.64 & -1.94 \\
0.35 & 2.31 & 2.22 & 2.92 & 2.00 \\
-2.94 & -0.01 & 4.31 & -2.51 & -0.19 \\
1.93 & -0.33 & -1.87 & -1.62 & 0.90
\end{pmatrix}$$

The Euclidean distances between the rows of this matrix are equal to those given in the matrix **D** shown in (3.1).   □

In the situation considered above, i.e. where the distance matrix, **D**, contains Euclidean distances derived from an $n \times p$ matrix, **X**, classical scaling can be shown to be equivalent to principal components analysis, with the derived coordinate values corresponding to the scores on the principal components found from the covariance matrix. This equivalence will be discussed in detail in Appendix A. One result of the duality of the techniques is that classical multidimensional scaling is also often referred to as *principal coordinates analysis* (see Gower, 1966b).

The more interesting and more practically relevant aspects of classical multidimensional scaling arise when considering its application to dissimilarity matrices, rather than sets of Euclidean distances calculated from multivariate data matrices.

First consider when the dissimilarity matrix of interest is Euclidean, as defined in Chapter 2. In such cases the matrix, **B**, will be positive semi-definite usually of rank $n - 1$ (the zero eigenvalue corresponding to the constraint on the centroid of the points), and an exact spatial representation of the observed dissimilarities, i.e. one for which $d_{ij} = \delta_{ij}$, can be found in an $(n - 1)$-dimensional Euclidean space. Since this clearly will not give a particularly parsimonious description of the dissimilarities, interest will now usually centre on finding suitable lower-dimensional configurations in which the distances between the points representing the individuals approximate the observed dissimilarities. It is easy to show that the procedure described previously for finding **X** already has the points referred to their principal axes so that the best fitting k $(k < n - 1)$ dimensional representation is given by the k eigenvectors of **B** corresponding to the k largest eigenvalues. The adequacy of the k-dimensional representation can be judged by the size of the following criterion

$$P_k = \frac{\sum_{i=1}^{k} \lambda_i}{\sum_{i=1}^{n-1} \lambda_i} \qquad (3.10)$$

Ideally, a value of k such that P_k is of the order of 80% would be sought (although see later).

Example 3.2

As an illustration of the use of classical MDS in the case of a Euclidean dissimilarity matrix, Gower's coefficient (see Section 2.2.3) will be calculated for the data on garden flowers shown in Table 3.1. There are no missing values in those data; consequently, the matrix of Gower's coefficients for each pair of flowers will be positive semi-definite, allowing a Euclidean representation to be found with distances $(1 - s_{ij})^{1/2}$. Because two plants with tubers have at least something in common, whereas plants without tubers may grow in completely different ways, co-absences of this variable are ignored in the calculation of the dissimilarity values, i.e. a weight of zero is used at the appropriate places in (2.4). The derived similarity matrix is shown in Table 3.2.

The non-zero eigenvalues of the matrix **B** for this example are shown in Table 3.3. Here the first six eigenvalues are required to give a value of the criterion

Table 3.1 Characteristics of some garden flowers. (Reproduced with permission from Kaufman and Rousseeuw, 1990.)

Flower	1	2	3	4	5	6	7
1. Begonia (*Bertinii bolivienis*)	0	1	1	4	3	25	15
2. Broom (*Cytisus praecox*)	1	0	0	2	1	150	50
3. Camellia (Japonica)	0	1	0	3	3	150	50
4. Dahlia (Tartini)	0	0	1	4	2	125	50
5. Forget-me-not (*Myosotis sylvatica*)	0	1	0	5	2	20	15
6. Fuchsia (Marinka)	0	1	0	4	3	50	40
7. Geranium (Rubin)	0	0	0	4	3	40	20
8. Gladiolus (Flowersong)	0	0	1	2	2	100	15
9. Heather (*Erica carnea*)	1	1	0	3	1	25	15
10. Hydrangea (Hortensis)	1	1	0	5	2	100	60
11. Iris (Versicolor)	1	1	1	5	3	45	10
12. Lily (*Lilium regale*)	1	1	1	1	2	90	25
13. Lily-of-the-valley (Convallaria)	1	1	0	1	2	20	10
14. Peony (*Paeonia lactiflora*)	1	1	1	4	2	80	30
15. Pink carnation (Dianthus)	1	0	0	3	2	40	20
16. Red rose (*Rosa rugosa*)	1	0	0	4	2	200	60
17. Scotch rose (*Rosa pimpinella*)	1	0	0	2	2	150	60
18. Tulip (*Tulipa sylvestris*)	0	0	1	2	1	25	10

Variables
1. May the plant be left in the garden when it freezes? (yes = 1, no = 0)
2. Can the plant be in shadow? (yes = 1, no = 0)
3. Is plant tuberous? (yes = 1, no = 0)
4. Colour of flowers (white = 1, yellow = 2, pink = 3, red = 4, blue = 5)
5. Type of soil in which plant thrives (1 = dry, 2 = normal, 3 = wet)
6. Height of plant (cm)
7. Ideal distance between plants (cm)

Table 3.2   Gower's similarity matrix for garden flower data

	1	2	3	4	5	6	7	8	9	10	11	12	13	14	15	16	17	18
1	1.00																	
2	0.09	1.00																
3	0.51	0.33	1.00															
4	0.53	0.41	0.41	1.00														
5	0.57	0.10	0.43	0.39	1.00													
6	0.77	0.21	0.71	0.48	0.56	1.00												
7	0.69	0.30	0.46	0.56	0.46	0.76	1.00											
8	0.51	0.43	0.29	0.74	0.51	0.32	0.51	1.00										
9	0.43	0.43	0.43	0.11	0.50	0.39	0.30	0.23	1.00									
10	0.24	0.42	0.42	0.38	0.61	0.39	0.14	0.30	0.45	1.00								
11	0.68	0.23	0.37	0.25	0.54	0.48	0.40	0.37	0.54	0.53	1.00							
12	0.49	0.31	0.31	0.47	0.49	0.35	0.23	0.53	0.49	0.61	0.64	1.00						
13	0.41	0.25	0.25	0.23	0.65	0.37	0.28	0.35	0.65	0.59	0.55	0.76	1.00					
14	0.63	0.32	0.32	0.62	0.48	0.52	0.37	0.51	0.48	0.61	0.63	0.83	0.61	1.00				
15	0.26	0.46	0.30	0.42	0.46	0.26	0.50	0.51	0.64	0.48	0.40	0.52	0.61	0.51	1.00			
16	0.16	0.59	0.25	0.63	0.18	0.29	0.39	0.36	0.19	0.57	0.16	0.38	0.33	0.53	0.55	1.00		
17	0.06	0.80	0.30	0.52	0.23	0.17	0.26	0.55	0.23	0.62	0.20	0.42	0.38	0.43	0.60	0.79	1.00	
18	0.56	0.50	0.21	0.52	0.41	0.32	0.53	0.78	0.41	0.08	0.41	0.33	0.28	0.33	0.39	0.15	0.33	1.00

Table 3.3   Non-zero eigenvalues of B for flower data

1	1.137
2	0.935
3	0.636
4	0.570
5	0.369
6	0.370
7	0.228
8	0.172
9	0.154
10	0.086
11	0.065
12	0.041
13	0.041
14	0.029
15	0.023
16	0.013
17	0.007

specified in (3.10) above 80%. A scatterplot matrix of the corresponding coordinate values is shown in Fig. 3.1. A trained botanist or a panelist on *Gardener's Question Time* may be able to give a convincing interpretation of this diagram, but for others, including the authors, unravelling its message is exceedingly difficult. Perhaps the requirements stated above for P_k to be above 80% is too demanding when choosing the appropriate number of dimensions, particularly in the light of the following comment made by Gnanadesikan and Wilk (1969);

Interpretability and simplicity are important in data analysis and any rigid inference of optimal dimensionality in the light of the observed values of a numerical index of goodness of fit, may not be productive.

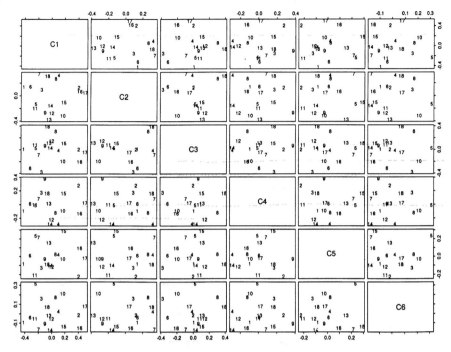

Figure 3.1 Scatterplot matrix of first six coordinate values from applying classical MDS to Gower's similarity matrix derived from the flower data in Table 3.1.

Two-dimensional solutions certainly have the virtue of simplicity and may, in many cases, provide an easily understood basis for gaining insight into complex proximity data; consequently they are, in many situations, likely to be the solutions of most practical importance. For the garden flower example, the first two coordinates give $P_2 = 43\%$, and the scatterplot formed from these two coordinates' values is shown in Fig. 3.2. Again interpretation is unfortunately not straightforward, although the first dimension might loosely be associated with the 'hardiness' of a plant and the second with the need for sunny or shady growing conditions. ☐

Any interpretation of Fig. 3.2 must keep in mind that the two-dimensional representation of the observed dissimilarities is only an approximation and there is always the possibility that the relationships between points in the scatterplot do not necessarily correspond accurately to the relationships between the corresponding flowers as implied by their similarity values. Methods for highlighting possible distortions will be discussed in the next chapter.

Finally, in this section let us consider what happens when the observed dissimilarity matrix is not Euclidean and consequently the matrix **B** is not positive semi-definite. In such cases some of the eigenvalues of **B** will be negative; correspondingly, some coordinate values will be complex. If, however, **B** has only a small number of small negative eigenvalues, a useful spatial

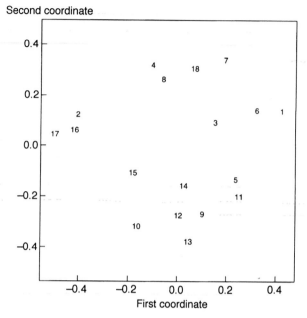

Figure 3.2   Scatterplot of first two coordinate values from applying classical MDS to Gower's similarity matrix derived from the flower data in Table 3.1.

representation of the observed dissimilarity matrix may often still result from the eigenvectors associated with the k largest positive eigenvalues. The adequacy of the resulting solution might be assessed using one or other of the following two criteria suggested by Mardia (1978)

$$P_k^{(1)} = \frac{\sum_{i=1}^{k} \lambda_i}{\sum_{i=1}^{n} |\lambda_i|} \qquad (3.11)$$

$$P_k^{(2)} = \frac{\sum_{i=1}^{k} \lambda_i^2}{\sum_{i=1}^{n} \lambda_i^2} \qquad (3.12)$$

Alternatively, Sibson (1979) recommends the following.

(1) *Trace criterion*: choose the number of coordinates so that the sum of the positive eigenvalues is approximately equal to the sum of all the eigenvalues.
(2) *Magnitude criterion*: accept as genuinely positive only those eigenvalues whose magnitude substantially exceeds that of the largest negative eigenvalue.

If, however, **B** has a considerable number of large negative eigenvalues, classical scaling of the dissimilarity matrix may be inadvisable and some other method of scaling (for example, non-metric scaling, see Section 3.4) might need to be considered. An alternative possibility is to transform **B** in a way that makes it positive semi-definite. Lingoes (1971) shows that such a transformation is achieved by adding a suitable constant to all off-diagonal elements of the symmetric dissimilarity matrix. Caillez (1983) shows how to calculate a

Table 3.4 First six eigenvectors obtained from classical MDS of flower data

Plant	1	2	3	4	5	6
1	0.42	0.14	−0.01	0.10	−0.08	0.09
2	−0.40	0.12	0.00	−0.27	−0.22	0.02
3	0.15	0.09	0.38	−0.17	−0.13	−0.16
4	−0.10	0.32	0.01	0.29	0.02	−0.02
5	0.24	−0.13	−0.02	0.01	0.23	−0.31
6	0.32	0.14	0.32	−0.01	0.01	0.04
7	0.19	0.34	0.09	−0.08	0.21	0.17
8	−0.06	0.26	−0.31	0.12	0.02	−0.17
9	0.10	−0.27	−0.06	−0.38	−0.03	0.11
10	−0.17	−0.32	0.20	0.09	−0.03	−0.21
11	0.25	−0.20	−0.08	−0.00	−0.21	0.05
12	0.00	−0.27	−0.14	0.22	−0.12	0.06
13	0.04	−0.38	−0.11	−0.01	0.15	0.02
14	0.02	−0.16	−0.03	0.29	−0.10	0.17
15	−0.18	−0.11	−0.11	−0.16	0.29	0.14
16	−0.42	0.06	0.21	0.13	0.09	0.15
17	−0.50	0.04	0.05	−0.02	−0.02	−0.08
18	0.07	0.31	−0.39	−0.16	−0.09	−0.07

suitable value for the required constant, but Mardia (1978) points out that the configuration obtained from this procedure will, in general, lead to a much poorer approximation than the configuration derived from the positive eigenvalues.

Example 3.3

As a first example of applying classical scaling to a non-Euclidean dissimilarity matrix (in fact a non-Euclidean distance matrix), consider the matrix of Manhattan distances (see Table 2.1) between the rows of the small data matrix given in Example 3.1

$$
\mathbf{D}_M = \begin{pmatrix}
0 \\
11 & 0 \\
18 & 11 & 0 \\
16 & 11 & 12 & 0 \\
6 & 11 & 18 & 18 & 0 \\
10 & 17 & 16 & 10 & 16 & 0 \\
11 & 16 & 15 & 7 & 15 & 9 & 0 \\
12 & 19 & 16 & 14 & 12 & 14 & 11 & 0 \\
15 & 18 & 13 & 17 & 13 & 13 & 16 & 11 & 0 \\
7 & 12 & 15 & 9 & 13 & 5 & 10 & 11 & 16 & 0
\end{pmatrix}
\tag{3.13}
$$

Applying classical MDS to $\mathbf{D}_M$ results in the following eigenvalues for $\mathbf{B}$

280.68, 249.42, 228.85, 92.51, 42.51, 21.97, 0.00, −15.07, −28.05, −56.68

Here, the first three positive eigenvalues dominate ($P_3^{(1)}$, for example, is 75%), and the corresponding eigenvectors could be used to represent the distances in $\mathbf{D}_M$.   □

Example 3.4

It is almost *de rigeur* in accounts of classical MDS that at least one example should involve a matrix of road distances between the major towns and cities in the writer's country of origin (see, for example, Manly, 1986; and Krzanowski and Marriot, 1994). Since the authors of this text are respectively, British (BSE) and German (SRH), application of a complex chance process was used to choose between a British and German road distances matrix, with the result shown in Table 3.5.

For fairly obvious reasons, the road distances would not be expected to be Euclidean. In fact the eigenvalues of the matrix $\mathbf{B}$ arising from the road distances include 12 negative values. The value of $P_2^{(1)}$ is 74% and a plot of the derived two-dimensional coordinates after reflection and rotation through 90° is shown in Fig. 3.3. This diagram gives a very good approximation to the map of Germany.

This example may be used to illustrate Sibson's recommendations on choice of dimensionality given earlier. The sum of the appropriate eigenvalues is 2 250 828; the sum of the first two positive eigenvalues is 2 262 576, so that using the trace criterion again leads to accepting $q = 2$ as the correct dimensionality for these data.   □

Least squares scaling

3.3 A fairly obvious approach to selecting a set of coordinate values with which to represent an observed dissimilarity matrix is by minimizing some least squares type fit criteria between the implicit distances and the dissimilarities. A simple measure of fit would be S_1 given by

$$S_1 = \sum_{i<j} (d_{ij}(\mathbf{x}_i, \mathbf{x}_j) - \delta_{ij})^2 \tag{3.14}$$

where $d_{ij}(\mathbf{x}_i, \mathbf{x}_j)$ represents the distance between the points with q dimensional coordinates $\mathbf{x}_i$ and $\mathbf{x}_j$, which represent stimuli i and j. (The problem of choosing an appropriate value for q will be discussed later.) In most applications d_{ij} is chosen to be Euclidean. The coordinate values that minimize the fit criterion, S_1 might be found by a standard optimization algorithm such as steepest descent, although in practice more 'tailored' algorithms are used (see later).

The use of the least squares goodness-of-fit function described above involves the following implicit assumption about the relationship between the fitted distances and the observed dissimilarities

$$d_{ij} = \delta_{ij} + \epsilon_{ij} \tag{3.15}$$

where the ϵ_{ij} represent a combination of errors of measurement and distortion errors arising because the dissimilarities may not correspond exactly to a configuration in q dimensions. In general, however, it may be more realistic

Table 3.5 Road distances between German cities

	1	2	3	4	5	6	7	8	9	10	11	12	13	14	15	16	17	18	19	20	21	22	23	24	25	26	27
1 Aachen (Aa)	–																										
2 Berlin (Be)	645	–																									
3 Bielefeld (Bi)	270	385	–																								
4 Bonn (Bo)	90	630	255	–																							
5 Bremen (Br)	415	390	155	375	–																						
6 Dortmund (Do)	150	485	110	140	255	–																					
7 Dresden (Dr)	645	195	500	555	485	530	–																				
8 Düsseldorf (Du)	80	565	185	75	345	70	610	–																			
9 Essen (Es)	115	530	150	105	295	35	560	35	–																		
10 Frankfurt/M (Fr)	260	545	275	175	460	310	455	230	255	–																	
11 Freiburg/Br (Fb)	535	815	595	445	735	560	730	500	525	280	–																
12 Hamburg (Ha)	505	290	235	490	120	345	485	425	390	510	785	–															
13 Hannover (Hn)	370	280	110	355	110	210	375	290	255	350	620	170	–														
14 Karlsruhe (Ka)	395	675	405	305	590	450	560	360	385	140	145	645	480	–													
15 Kassel (Ks)	325	365	120	295	270	175	350	245	210	190	470	320	155	330	–												
16 Kiel (Ki)	600	355	340	585	215	440	550	520	465	605	880	95	265	740	415	–											
17 Köln (Ko)	70	600	220	25	370	115	575	45	70	190	465	460	325	320	320	550	–										
18 Leipzig (Le)	570	185	370	485	370	410	115	540	440	380	655	395	260	510	215	415	500	–									
19 Mainz (Mz)	245	570	343	160	462	236	500	215	242	36	269	531	365	141	219	632	170	395	–								
20 Mannheim (Mn)	330	615	615	245	530	385	525	300	325	75	205	580	420	65	270	675	260	450	71	–							
21 München (Mc)	670	585	520	580	750	695	460	635	660	415	340	795	630	285	500	890	595	425	384	340	–						
22 Münster (Mu)	205	460	80	210	175	60	585	130	90	355	620	295	185	470	170	415	170	420	295	415	750	–					
23 Nürnberg (Nu)	485	435	455	395	585	510	305	450	475	225	380	645	470	240	240	740	170	415	255	165	165	590	–				
24 Saarbrücken (Sa)	280	740	430	245	655	360	650	295	325	205	285	705	540	145	395	800	255	575	154	135	430	430	390	–			
25 Stuttgart (St)	465	625	525	375	665	490	495	430	450	220	120	715	550	80	400	810	395	460	203	135	215	560	190	225	–		
26 Würzburg (Wu)	380	495	350	295	490	410	370	350	375	120	335	545	380	195	230	640	310	335	151	160	270	415	105	300	165	–	
27 Wuppertal (Wp)	110	525	150	85	295	40	565	30	35	245	515	385	250	375	215	480	60	445	200	315	650	95	465	315	450	355	–

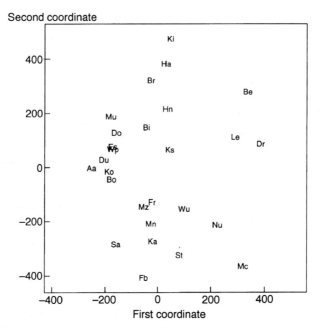

Figure 3.3   Scatterplot of first two coordinate values from applying classical MDS to German road distances (Table 3.5).

to postulate a rather less simple relationship between the dissimilarities and distances, for example

$$d_{ij} = \alpha + \beta\delta_{ij} + \epsilon_{ij} \tag{3.16}$$

Assuming such a relationship, minimization of the goodness-of-fit criterion described above would now involve an essentially two-stage process. First, for a given q-dimensional configuration, $x_1, x_2, \ldots, x_n$, and hence a given set of distances, estimates of α and β would be obtained from a simple linear regression of d_{ij} on the δ_{ij}. This leads to a set of estimated distances, $\hat{d}_{ij}$, more generally known in a scaling context as *disparities*, calculated from

$$\hat{d}_{ij} = \hat{\alpha} + \hat{\beta}\delta_{ij} \tag{3.17}$$

Now some optimization algorithm, such as steepest descent, would be applied to find a revised configuration (i.e. a set of coordinate values), minimizing the following goodness-of-fit criterion

$$\sum_{i<j} (d_{ij} - \hat{d}_{ij})^2 \tag{3.18}$$

Extending these ideas to allow the distances and dissimilarities to be related by a more complex relationship than the one specified in (3.16), leads to the following possible goodness-of-fit criterion for least squares scaling

$$S_2 = \sum_{i<j} (d_{ij} - f(\delta_{ij}))^2 \tag{3.19}$$

where f represents the functional relationship assumed between distances and dissimilarities, and $f(\delta_{ij})$ represents the estimated disparity value corresponding to δ_{ij} calculated from fitting this relationship. Again this estimated value will usually be referred to as $\hat{d}_{ij}$.

The least squares fit criterion in (3.19) is invariant under rigid transformations of the derived spatial representation, i.e. under rotations, translations and reflections; this is clearly desirable (although see Section 3.5). Unfortunately the criterion is not invariant under non-rigid transformations such as uniform stretching and shrinking. In other words, if a configuration defined by $x_1, x_2, \ldots, x_n$, is amended to $kx_1, kx_2, \ldots, kx_n$, where k is a constant, and the disparities are scaled accordingly, then the value of S_2 will change; this is clearly undesirable because simply enlarging $(k > 1)$ or shrinking $(k < 1)$ a configuration should not alter the measurement of how well it fits the dissimilarities, since the relationships between the distances do not change. As a result, a scaling factor is needed which has the same dependence on the scale of the configuration as does S_2. The factor $\sum_{i<j} d_{ij}^2$ has the required property, leading to the revised fit criterion, S_3

$$S_3 = \frac{\sum_{i<j}(d_{ij} - f(\delta_{ij}))^2}{\sum_{i<j} d_{ij}^2} \tag{3.20}$$

The criterion S_3 has all the desirable properties of S_2, but in addition is invariant under changes of scale, i.e. uniform stretching or shrinking. (Other possible normalizing factors are considered in Coxon, 1982b).

A further amendment to the fit criterion that is often considered is the introduction of 'weights' into its numerator to allow the derived spatial representation to be biased towards a more accurate portrayal of particular dissimilarities. The revised fit criterion, S_4 is now

$$S_4 = \frac{\sum_{i<j} w_{ij}(d_{ij} - f(\delta_{ij}))^2}{\sum_{i<j} d_{ij}^2} \tag{3.21}$$

The weights can be chosen for specific purposes; if, for example, it was desired, that small observed dissimilarities be represented more accurately than larger values, then w_{ij} could be set equal to δ_{ij}^{-1}. When the weights in S_4 are all unity, its square root is generally referred to in the context of MDS as *stress*, although this term is also used to refer to a number of other similar fit criteria, as will be seen in later chapters.

(This is a convenient point to note that classical multidimensional scaling can also be formulated in terms of the optimization of a particular goodness-of-fit function, as will be described in Appendix A.)

Example 3.5

As an example of the approach to scaling described above it will be applied to the judgements of the dissimilarities of a number of World leaders and politicians prominent at the time of the Second World War. A number of subjects made ratings on a nine-point scale ranging from 1, indicating very similar, to 9, indicating very dissimilar. The ratings of one of the subjects are shown in Table 3.6. The two-dimensional solution obtained from minimizing

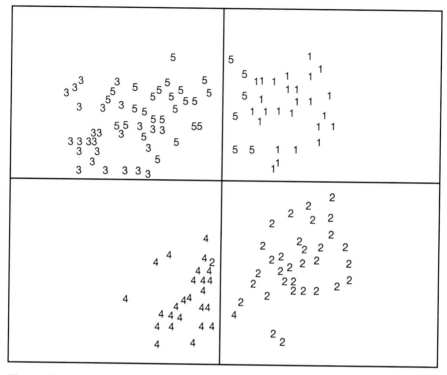

Figure 3.6   Data generated to contain five groups displayed in the space of the first two principal components.

solution corresponding to the lowest value of the fit criterion. A similar problem occurs in all applications of multidimensional scaling methods which operate by minimizing a goodness-of-fit criterion.

Non-metric multidimensional scaling

3.4   In both classical and least squares scaling, goodness-of-fit measures are all based on direct numerical comparison of dissimilarities (usually after some transformation) and distances, and disparities are obtained by metric regression methods. Consequently, both procedures are often referred to as methods for metric scaling. In many situations, however, it might be believed that the observed proximities contain little reliable information beyond their rank order, i.e. the proximities do not have strict numerical significance and so assuming say, a linear or quadratic relationship between the fitted distances and the observed proximities, will no longer be appropriate. In psychological experiments, for example, proximity matrices frequently arise from asking human subjects to make judgements about the similarity or dissimilarity of stimuli of interest; in many such experiments the investigator may feel that, realistically, subjects can only give 'ordinal' judgements, e.g. in comparing a range of colours they might be able to specify that one was 'brighter' than another without being able to attach any value to the exact numerical difference between them. Alternatively subjects are able to say that one pair of stimuli are more alike than another pair etc.

For such situations what is needed is a method of multidimensional scaling, the solutions from which depend only on the rank order of the proximities, rather than on their actual numerical values. In other words, the method should be invariant under monotonic transformation of the proximity matrix. A major breakthrough in multidimensional scaling was achieved in the early 1960s by Shepard (1962a,b) and Kruskal (1964a) when such a method was eventually derived. The quintessential component of the method is that the disparities are now calculated using what is known as *monotonic regression* (see Barlow *et al.*, 1972). The details of the method are described later, but in essence its aim is to represent the fitted distances as

$$d_{ij} = \hat{d}_{ij} + \epsilon_{ij} \tag{3.22}$$

where the disparities, $\hat{d}_{ij}$, are monotonic with the observed dissimilarities, δ_{ij}, and, subject to this constraint, resemble the d_{ij} as closely as possible. If the observed dissimilarities are ranked from lowest to highest as

$$\delta_{i_1,j_1} < \delta_{i_2,j_2} < \cdots < \delta_{i_N,j_N} \tag{3.23}$$

where $N = n(n-1)/2$, then the $\hat{d}_{ij}$ can be chosen to satisfy either the weak monotonicity (WM) or strong monotonicity (SM) conditions

$$\text{WM}: \hat{d}_{i_1,j_1} \leqslant \hat{d}_{i_2,j_2} \leqslant \cdots \leqslant \hat{d}_{i_N,j_N} \tag{3.24}$$

$$\text{SM}: \hat{d}_{i_1,j_1} < \hat{d}_{i_2,j_2} < \cdots < \hat{d}_{i_N,j_N} \tag{3.25}$$

In practice the first condition is generally used, since requiring strong monotonicity places more restrictions on the configuration and consequently leads to larger stress values.

In many situations there will be ties amongst the observed dissimilarities, i.e. $\delta_{rs} = \delta_{tu}$. In the so-called *primary approach* to such ties the corresponding disparity values are not required to equal each other; in the *secondary approach*, equality of disparity values is required. The latter procedure has been shown by many authors, for example, Kendall (1971) and Lingoes and Roskam (1973), to be very restrictive and generally less satisfactory than the primary approach.

The monotonic regression procedure for determining the required disparities involves an explicit minimization of a function measuring the departure from perfect monotonicity between dissimilarities and fitted distances. In general, a least squares type function is employed leading to least squares monotonic regression which is described in detail by Kruskal (1964b) and Bartholomew (1959). The following small example, taken from Gordon (1981) serves to illustrate the essential features of the method.

Assume that **a**, **b** and **c** are three sequences, each containing m real numbers, $(a_1, a_2, \ldots, a_m)$, $(b_1, b_2, \ldots, b_m)$ and $(c_1, c_2, \ldots, c_m)$ respectively. (In the context of the discussion in Section 3.3, let **a** be identified with the observed dissimilarities, δ_{ij}, **b** with the fitted distances, d_{ij}, and **c** with the disparities, $\hat{d}_{ij}$.)

Now suppose specifically that $\mathbf{a} = (1, 2, 4, 4, 6, 8, 9, 10, 11, 15)$ and $\mathbf{b} = (1, 4, 5, 6, 7, 8, 12, 13, 13, 14)$, and we require to find values in **c** that satisfy

$c_k \leqslant c_l$ whenever $a_k \leqslant a_l$ and resemble the values in **b** as closely as possible. If the primary approach is adopted to ties then choosing $c_2 = b_2$, $k = 1 \ldots 10$ ensures a perfect resemblance between **c** and **b** with **c** satisfying the required monotonic requirement. If the secondary approach to ties is adopted then since $a_3 = a_4$ we require $c_3 = c_4$ and a perfect resemblance between **c** and **b** is not possible. Here the values in **c** are chosen to minimize the sum of squares criterion

$$S^*(\mathbf{c}) = \sum_{k=1}^{m} (b_k - c_k)^2 \qquad (3.26)$$

In this example it is easy to see that the solution that minimizes S^* is one in which $c_k = b_k$, $k = 1, 2, 5, 6, \ldots, 10$ and $c_3 = c_4 = (b_3 + b_4)/2$. In the general case, **c** will consist of blocks containing elements with consecutive indices, for example $(c_{r+1}, c_{r+2}, \ldots, c_s)$, with each element in a block equalling the mean of the corresponding set of values in **b** – in this case

$$\sum_{k=r+1}^{s} b_k/(s - r) \qquad (3.27)$$

and such that the common values increase from block to block.

Although the regression procedure used to derive disparities in this method of scaling is more involved than for the least squares scaling methods described in the previous section, finding the coordinate values that optimize stress is essentially the same problem and can be attacked using similar optimization algorithms. Several such algorithms have been suggested (see, for example, Kruskal, 1964b and Kruskal *et al.*, 1977), most based on the method of steepest descent, although the *majorization algorithm* of de Leeuw and Heiser (1977) has also been used (see de Leeuw, 1988).

A problem with all algorithms is that of local minima, and the usual advice given to help overcome the problem is to repeat the analyses from a number of different initial configurations. A detailed investigation of the local minima problem is reported in Groenen and Heiser (1996), who show that unidimensional scaling is particularly prone to local minima, whereas full dimensional scaling, i.e. with dimensionality $n - 1$, and with Euclidean distances, has only a global minimum. For intermediate numbers of dimensions with Euclidean distances, the severity of the local minima problem depends on the observed proximities. Groenen and Heiser (1996) develop a procedure which alternates a local search step, in which a local minimum is sought, with a 'tunnelling step' in which a different configuration is sought with the same stress as the previous local minimum. In this manner successively better local minima are obtained, and Groenen and Heiser's experiments indicate that, in many cases, the last of these is the required global minimum.

In some situations it might be required to apply non-metric scaling to sets of proximities in which not all the pairwise values have been observed. If, for example, the number of stimuli is large, the experimenter may decide in advance to observe only a fraction of the possible proximities for reasons of cost. The method can be adapted to such a situation by simply requiring the summations in the definition of stress (3.21), to run over those pairs i and j for which δ_{ij} is observed, although there are times when such an approach can break down. One identified by Kruskal (1964b) is when the stimuli are split into two groups and the only dissimilarities observed are those between stimuli in different groups. In this case there is a simple zero-stress configuration in one dimension, namely two distinct points, where each point represents all the stimuli in one group.

(The scaling technique described above is almost universally referred to as non-metric scaling, since it is not assumed that the observed dissimilarities contain any metric information; all that is used is their rank order. Gordon, 1981, however, makes the point that the resulting geometrical configuration of points does contain metric information, and that consequently a more appropriate name might be *ordinal scaling*.)

As with the methods of scaling described in Sections 3.2 and 3.3, in the practical application of non-metric scaling an important question concerns the appropriate dimensionality of the spatial configuration needed to represent adequately a set of observed proximities. The points made about simplicity and interpretability in Section 3.2 apply equally well to the solutions from non-metric scaling, so any rigid interpretation of a numerical index purporting to indicate the correct dimensionality remains bad practice. Nevertheless, there are some relatively informal procedures that may be useful in particular situations. Kruskal (1964a), for example, on the basis of his own experience with experimental and synthetic data, gives the following verbal evaluation of various values of stress.

Stress (%)	Goodness-of-fit
20	Poor
10	Fair
5	Good
2.5	Excellent
0	Perfect

'Scree' type diagrams, in which stress is plotted against dimensionality, are also occasionally useful, with the appropriate value of q corresponding to a pronounced 'elbow' in the diagram.

Wagenaar and Padmos (1971) indicate that the interpretation of stress is strongly dependent on the number of stimuli involved and thus conclude that a simple interpretation in terms of Kruskal's suggestion given above is often not justified. Spence (1970, 1972) and Spence and Graef (1974) report on an extensive set of Monte Carlo experiments, the results of which are purported to allow a more objective assessment to be made of underlying dimensionality. The simulated data were generated for a wide range of conditions similar to those that might be experienced by typical users. The number of stimuli was varied from 12 to 36, spaces of true dimensionality from one to four were investigated, and the level of error in the data varied from zero to infinity. Observed values of stress might be compared with the stress values generated in these experiments in order to decide on the 'true' dimensionality of a proximity matrix (see Kruskal and Wish, 1978, for an example).

Example 3.6

As a first example of the application of non-metric scaling, the data in Table 3.6 will again be used. The two-dimensional solution in this case is shown in Fig. 3.7.

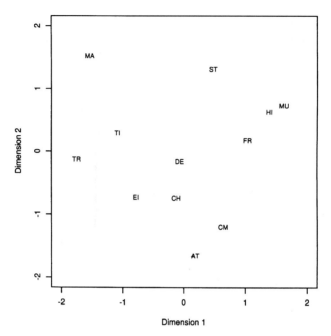

Figure 3.7   Two-dimensional solution from non-metric scaling of dissimilarities of World War II politicians.

The solution is clearly very similar to that achieved by metric scaling (see Fig. 3.4), although the stress value here is a little lower, being 0.20, and the squared multiple correlation coefficient is 0.74. The change from the initial configuration with stress value 0.22 to the final solution is shown in Fig. 3.8.

Often useful in the assessment of a scaling solution are plots of the observed dissimilarities against the distances from the scaling solution and of the disparities against the distances (such plots are often referred to as *Shepard diagrams*). The two plots for this example are shown in Figs 3.9 and 3.10. The second of these plots shows the large deviation from monotonicity of the disparities (which remember *are* monotonic with the observed dissimilarities) and the distances corresponding to the scaling solution. The scaling solution is only partially successful in this case.   ☐

Example 3.7

As a further more substantial example of non-metric scaling, the method will be applied to data arising from an investigation into the taxonomic status of British water voles, reported in Corbet *et al.* (1970). The original data consisted of recordings of the presence or absence of 13 characteristics in about 300 water vole skulls divided into samples from 14 populations from Britain and the rest of Europe. These data were used to construct a matrix of dissimilarities between the 14 populations; this matrix is shown in Table 3.7. (Details of how the dissimilarities were calculated are given in the original paper.) Non-metric

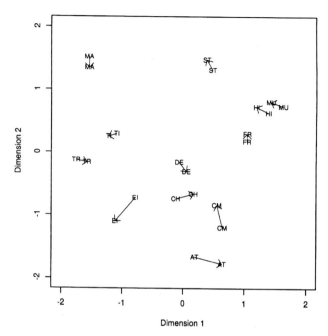

Figure 3.8   Change in two-dimensional coordinates from initial to final solution in non-metric scaling of World War II politicians.

scaling solutions were obtained for one to four dimensions. The resulting stress and R^2 values were as follows.

Dimensions	Stress	Multiple R^2
1	0.29	0.78
2	0.15	0.90
3	0.08	0.96
4	0.04	0.99

The two-dimensional solution is shown in Fig. 3.11. The diagram suggests an apparent division of the populations into three main groups. The first, Pyrenees II, North Spain, South Spain, are very clearly separated from the remainder. The six British populations appear to form a group with the German and Norwegian populations, with Alps, Yugoslavia, Pyrenees I constituting a third group. The stress value associated with this solution is, however, only in the 'fair' to 'poor' range, and in Chapter 4 we shall return to this example to investigate the possible distortions produced by mapping these data into two dimensions.   □

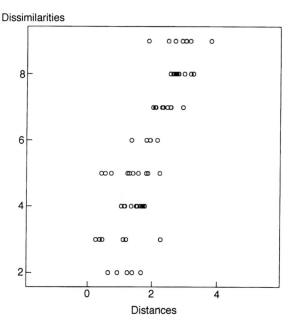

Figure 3.9 Plot of observed dissimilarities against distances for two-dimensional non-metric scaling solution of World War II politicians.

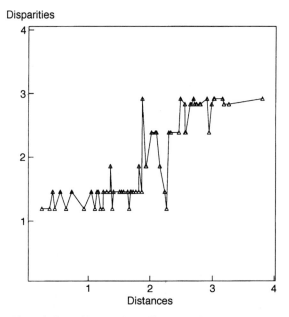

Figure 3.10 Plot of disparities against distances for two-dimensional non-metric scaling solutions of World War II politicians.

Table 3.7 Matrix of dissimilarities for 14 populations of water voles

	1	2	3	4	5	6	7	8	9	10	11	12	13
2	0.099												
3	0.033	0.022											
4	0.183	0.114	0.042										
5	0.148	0.224	0.059	0.068									
6	0.198	0.039	0.053	0.085	0.051								
7	0.462	0.266	0.322	0.435	0.268	0.025							
8	0.628	0.442	0.444	0.406	0.240	0.129	0.014						
9	0.113	0.070	0.046	0.047	0.034	0.002	0.106	0.129					
10	0.173	0.119	0.162	0.331	0.177	0.039	0.089	0.237	0.071				
11	0.434	0.419	0.339	0.505	0.469	0.390	0.315	0.349	0.151	0.430			
12	0.762	0.633	0.781	0.700	0.758	0.625	0.469	0.618	0.440	0.538	0.607		
13	0.530	0.389	0.482	0.579	0.597	0.498	0.374	0.562	0.247	0.383	0.387	0.084	
14	0.586	0.435	0.550	0.530	0.552	0.509	0.369	0.471	0.234	0.346	0.456	0.090	0.038

Populations
1: Surrey; 2: Shropshire; 3: Yorkshire; 4: Perthshire; 5: Aberdeen; 6: Eilean Gamhna; 7: Alps; 8: Yugoslavia; 9: Germany; 10: Norway; 11: Pyrenees I; 12: Pyrenees II; 13: North Spain; 14: South Spain.

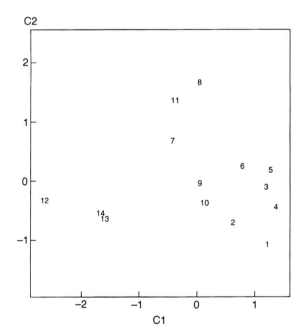

Figure 3.11 Scatterplot matrix of two-dimensional non-metric scaling solution of water vole population dissimilarities.

Example 3.8

MDS methods have, in the past, been primarily used in ecology, psychology and related disciplines. An interesting recent account of an application in a totally different area, namely epidemiology, is given in Cliff *et al.* (1995). The example involves measles mortality data for six states in Australia and for New Zealand between 1860 and 1949. The matrix of intercorrelations between the seven time

Table 3.8   Australia and New Zealand, 1860–1949; Pearson correlation coefficients between state time series of measles mortality

State	NSW	Vic	Qld	SA	WA	Tas	NZ
New South Wales (NSW)	1.000						
Victoria (Vic)	0.809	1.000					
Queensland (Qld)	0.773	0.666	1.000				
South Australia (SA)	0.431	0.433	0.337	1.000			
Western Australia (WA)	0.118	0.134	0.181	0.320	1.000		
Tasmania (Tas)	0.751	0.893	0.650	0.267	0.040	1.000	
New Zealand (NZ)	0.700	0.628	0.695	0.600	0.119	0.567	1.000

series is shown in Table 3.8. The two-dimensional classical scaling solution for the dissimilarity matrix obtained by taking $(1 - r)^{1/2}$ is shown in Fig. 3.12. This is similar to the solution from non-metric scaling reported by Cliff *et al.* (1995) although New Zealand is further away from the NSW, Queensland, Victoria and Tasmania cluster. Cliff *et al.* (1995) gives details of further scaling solutions used to study temporal changes in the patterns of measles mortality in the different regions.   □

Other medical applications of the method are given in Goldenberger *et al.* (1989) and Pollock and Cliff (1992).

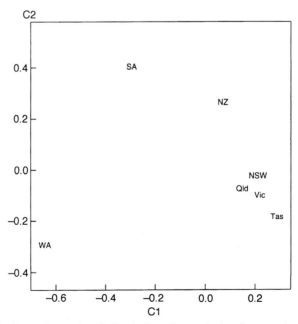

Figure 3.12   Two-dimensional classical scaling solution for measles data.

Non-Euclidean configurations

3.5 In most applications of MDS the distance function assumed for the spatial configuration to be derived is Euclidean. In principle, however, there is no reason why the definition of goodness-of-fit criteria used in least squares and non-metric scaling could not be used with other distance functions, particularly the Minkowski metrics introduced in Chapter 2. Kruskal (1964a), for example, describes the application of non-metric scaling to judged similarities of 14 spectral colours (Ekman, 1954), using a variety of Minkowski metrics. In each case the familiar colour circle was obtained, although the precise shape, spacing, and angular correlation varies with the particular metric employed. Figure 3.13 shows the stress of the best-fitting configuration as a function of r in the Minkowski metric definition given in Table 2.1. A value of 2.5 for r gives the best fit, and there is some indication that subjective distance between colours may be slightly non-Euclidean.

All Minkowski metrics possess invariance under translations and under central reflections of the coordinate axes. Only in the Euclidean case, however,

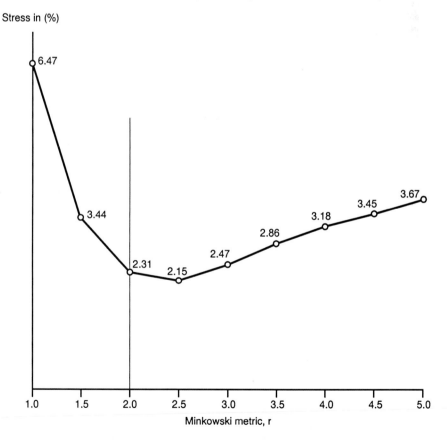

Figure 3.13 Stress as a function of r (reproduced with permission from Ekman, 1954).

are distances invariant under orthogonal rotations, and such invariance is used by Attneave (1950) as one of the arguments against assuming that 'psychological space' is Euclidean in character. According to Attneave;

> This latent hypothesis implies that if two stimuli in a two-dimensional space differ by a units on one axis and by b units on another, their total difference is given by the Pythagorean theorem, and is equal to $(a^2 + b^2)^{1/2}$. Its most important psychological implication is that, within a space of given dimensionality, one set of axes or one frame of reference is as good as any other. For example, colours are commonly described in terms of 'brightness' and 'saturation'. If psychological space were Euclidean, it would be entirely possible and psychologically proper to rotate these axes and discuss colours in terms of attributes which might be termed 'sightness' and 'braturation', axes cutting across those we generally use.

Attneave then proceeds to make the following case for the city block metric for psychological space

> The psychological implication is that there is a unique coordinate system in psychological space, upon which 'distances' between stimuli are strictly dependent; and thus the choice of axes is to be dictated, not by linguistic expendiency, but by psychological fact.

Attneave's arguments are taken up in more detail in the context of multidimensional scaling by Arabie (1991), who provides the following quotation from Shepard (1980) which gives reasons why the Manhattan metric might be a more suitable choice for describing psychological space.

> This question of the form of the metric of perceptual space might seem to be rather esoteric and of little consequence for major issues of human behaviour. I believe, however, that there is a fundamental connection between the sharp-cornered form of the isosimilarity contour of the city block metric and some of the significant phenomena of discontinuous or insightful learning and of instabilities of choice...

The isosimilarity contours mentioned in the quotation from Shepard are shown in Fig. 3.14 both for city-block and Euclidean distance. The Euclidean isosimilarity contours are circles (in two dimensions) or ellipses, if the dimensions are unequally weighted – represented as stretching or shrinking of the axes. Because this curve is continuous, a small change in the relative weighting of the two dimensions results in only a small shift in the locus of points around a given, 'standard' stimulus. But the diamond-shaped isosimilarity contours of the two-dimensional city-block spaces afford discontinuity: a small change in the weighting of the two dimensions can cause major shifts in judged similarity.

More discussion of this issue is presented in Cross (1965a, 1965b), Carroll and Wish (1974), and Eisler and Lindman (1990).

Summary

3.6   There are a variety of methods for obtaining a spatial representation of proximity data. Classical multidimensional scaling is, in many respects, closely related to principal components analysis (see Appendix A), but can be applied to distances and dissimilarities arising directly rather than simply derived from a multivariate data matrix. The method can also be applied to non-Euclidean distance matrices although the question of whether the solution given is

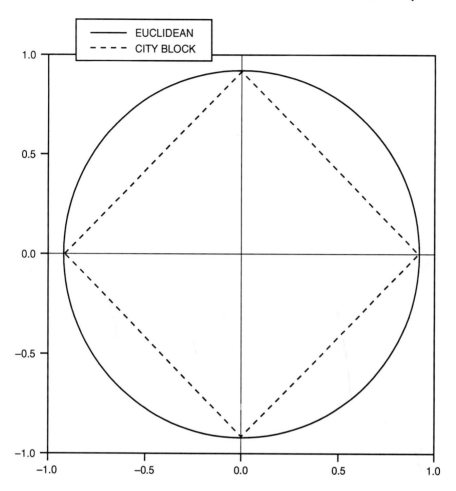

Figure 3.14 Isosimilarity contours for Euclidean and Manhattan distances.

acceptable depends on the relative magnitudes of the positive and negative eigenvalues of the inner products matrix, **B**. Metric and non-metric scaling will often give similar solutions when applied to the same set of proximities, although those given by the latter will generally have lower stress values.

In most situations, of course, a representation obtained by any of these scaling methods should not be regarded merely as an end in itself. The primary purpose of such a representation is to enable the investigator to gain a better understanding of the total underlying pattern of interrelations in the data, and hence to decide what further observations, experiments, or modifications to theory will most advance the science as a whole.

4

Interpreting, Diagnosing and Comparing Multidimensional Scaling Solutions

Introduction

4.1 Having arrived at a possible multidimensional scaling solution it is important to consider how to interpret the solution and whether the derived configuration reflects accurately the relationships between the stimuli implied by the observed proximities. In particular the investigator needs to assess whether the 'mapping' process has resulted in any significant distortion of these relationships. Unless the stress value of a solution is zero or very close to zero, some misrepresentation of the stimuli relative to each other is inevitable, and it becomes important to investigate where distortions arise and which stimuli are most poorly represented in the scaling solution.

An additional problem considered in this chapter is that of comparing alternative multidimensional scaling solutions. Suppose, for example, that a number of researchers have collected judgements of similarity on the same set of stimuli and subjected their data to one or other method of multidimensional scaling. In this way, each investigator will produce a particular 'map' of the stimuli and the question immediately arises, 'are the maps the same?', or perhaps more realistically, 'what is the degree of correspondence of the alternative solutions?'

Interpreting solutions

4.2 Once a multidimensional scaling configuration for a set of observed dissimilarities or similarities has been found, the question of interpretation arises. In general this means trying to arrive at some sensible explanation of the configuration in terms of labelling the axes. In many cases labels are suggested on the basis of simply looking at the configuration and recalling what is known about the stimuli. This method is, however, rather limited and also prone to the tendency to find 'patterns' whether or not they exist. One difficulty is that investigators sometimes forget that (at least for Euclidean distances) *all* directions in the configuration need to be examined, not only the directions of the coordinate axes.

A useful alternative procedure for interpreting a MDS configuration is to regress the values of some variable associated with the stimuli and suspected to have a systematic relationship to position in the scaling configuration, on the derived coordinate values. If the regression is reasonably successful as judged by, say, the value of the multiple correlation coefficient, then an axis for the variable can be defined through the origin of the configuration using the direction cosines

$$\frac{\hat{\beta}_i}{(\sum_{i=1}^{q} \hat{\beta}_i^2)^{1/2}} \tag{4.1}$$

(see also the external vector model described in Chapter 5).

Example 4.1

As an example, consider the data shown in Table 4.1. These data relate to air pollution in 41 US cities. Air pollution is measured by the annual mean concentration of sulphur dioxide, in micrograms per cubic metre. The other six variables relate to human ecology and climate. Here, the standardized variable values, apart from air pollution, were used to calculate a matrix of Manhattan distances between the cities and the resulting distance matrix was subjected to classical MDS. The configuration corresponding to the first two derived coordinates is shown in Fig. 4.1. (The value of $P_2^{(1)}$ for this solution is 49%.) Regressing air pollution on the two coordinate values for each city gives the following results

$$\hat{\beta}_0 = 30.05, \quad \hat{\beta}_1 = -4.70, \quad \hat{\beta}_2 = 1.024$$

The multiple correlation coefficient is 0.50. This is not particularly impressive, but does indicate some relationship between the MDS coordinates, particularly the first, and air pollution. Figure 4.2 shows the axis defined by the regression; it is largely the same as the first scaling axis.

Figure 4.1 has two very clear outliers, cities 1 (Phoenix) and 11 (Chicago); the first has low pollution, the second very high pollution. It is left as an exercise for the reader to investigate whether removal of these cities causes any dramatic change in the scaling solution and its interpretation. □

In addition to looking for interpretable axes, there are a number of other ways that a multidimensional scaling solution might be examined in the search to uncover its message. The way points cluster into groups, for example, may be informative in many circumstances (see Example 3.7). In some situations more exotic structures might be visible – the famous colour circle (Shepard, 1962b), for example, or the tetrahedron structure in three dimensions reported by Künnapas *et al.* (1964) for similarity judgements of geometrical figures, with triangles, circles, squares and crosses on the four vertices. But perhaps the most well known of these is the horseshoe shape, frequently seen in examples where the objective of multidimensional scaling is to order a set of stimuli according to an underlying attribute that is not directly observable but is believed to contribute substantially to the observed dissimilarities. This is usually referred

Table 4.1 Air pollution data for cities in the USA

City	1	2	3	4	5	6	7
1 Phoenix	10	70.3	213	582	6.0	7.05	36
2 Little Rock	13	61.0	91	132	8.2	48.52	100
3 San Francisco	12	56.7	453	716	8.7	20.66	67
4 Denver	17	51.9	454	515	9.0	12.95	86
5 Hartford	56	49.1	412	158	9.0	43.37	127
6 Wilmington	36	54.0	80	80	9.0	40.25	114
7 Washington	29	57.3	434	757	9.3	38.89	111
8 Jacksonville	14	68.4	136	529	8.8	54.47	116
9 Miami	10	75.5	207	335	9.0	59.80	128
10 Atlanta	24	61.5	368	497	9.1	48.34	115
11 Chicago	110	50.6	3344	3369	10.4	34.44	122
12 Indianapolis	28	52.3	361	746	9.7	38.74	121
13 Des Moines	17	49.0	104	201	11.2	30.85	103
14 Wichita	8	56.6	125	277	12.7	30.58	82
15 Louisville	30	55.6	291	593	8.3	43.11	123
16 New Orleans	9	68.3	204	361	8.4	56.77	113
17 Baltimore	47	55.0	625	905	9.6	41.31	111
18 Detroit	35	49.9	1064	1513	10.1	30.96	129
19 Minneapolis	29	43.5	699	744	10.6	25.94	137
20 Kansas City	14	54.5	381	507	10.0	37.00	99
21 St. Louis	56	55.9	775	622	9.5	35.89	105
22 Omaha	14	51.5	181	347	10.9	30.18	98
23 Albuquerque	11	56.8	46	244	8.9	7.77	58
24 Albany	46	47.6	44	116	8.8	33.36	135
25 Buffalo	11	47.1	391	463	12.4	36.11	166
26 Cincinnatti	23	54.0	462	453	7.1	39.04	132
27 Cleveland	65	49.7	1007	751	10.9	34.99	155
28 Columbus	26	51.5	266	540	8.6	37.01	134
29 Philadelphia	69	54.6	1692	1950	9.6	39.93	115
30 Pittsburg	61	50.4	347	520	9.4	36.22	147
31 Providence	94	50.0	343	179	10.6	42.75	125
32 Memphis	10	61.6	337	624	9.2	49.10	105
33 Nashville	18	59.4	275	448	7.9	46.00	119
34 Dallas	9	66.2	641	844	10.9	35.94	78
35 Houston	10	68.9	721	1233	10.8	48.19	103
36 Salt Lake City	28	51.0	137	176	8.7	15.17	89
37 Norfolk	31	59.3	96	308	10.6	44.68	116
38 Richmond	26	57.8	197	299	7.6	42.59	115
39 Seattle	29	51.1	379	531	9.4	38.79	164
40 Charleston	31	55.2	35	71	6.5	40.75	148
41 Milwaukee	16	45.7	569	717	11.8	29.07	123

Variables
1. SO_2 content of air in micrograms per cubic metre
2. Average annual temperature in °F
3. Number of manufacturing enterprises employing 20 or more workers
4. Population size in thousands
5. Average wind speed in miles per hour
6. Average annual precipitation in inches
7. Average number of days with precipitation per year

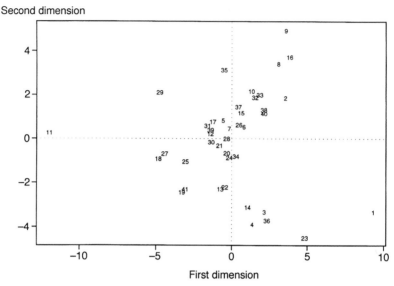

Figure 4.1 Two-dimensional classical MDS solution from Manhattan distances between 41 US cities.

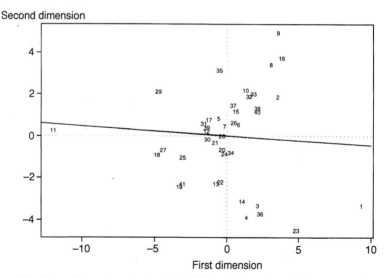

Figure 4.2 Two-dimensional classical MDS solution from Manhattan distances between 41 US cities showing an axis corresponding to regressing air pollution on two coordinate values.

to as a *seriation* problem and most common are examples in which the underlying attribute is time and the derived positions of the stimuli suggest their chronology. An example given by Kendall (1970) postulates a similarity matrix for a set of archeological graves, the elements of which are simply the number of varieties of artefacts shared by each pair of graves. The number of varieties in common decreases as the time gap between pairs of graves increases and a scaling solution might be expected to show a linear ordering of the graves according to age. Time intervals above some threshold, however, are typically indistinguishable from each other because, from that threshold onwards, pairs of graves have no artefact varieties in common. Therefore, if the graves are ordered from earliest to most recent, the similarity matrix contains non-zero elements only in a narrow band around the leading diagonal.

Example 4.2

To illustrate his archeological graves example, Kendall (1970) created an artificial similarity matrix for 51 hypothetical graves in chronological order, separated by equal time-intervals as follows

$$
s_{ij} = \begin{cases} 9 & \text{if } i=j \\ 8 & \text{if } 1 \leqslant |i-j| \leqslant 3 \\ \vdots \\ 1 & \text{if } 2 \leqslant |i-j| \leqslant 2 \\ 0 & \text{if } |i-j| > 25 \end{cases}
$$

Note that all pairs of graves separated by between 25 and 50 time intervals are lumped together.

Converting the generated similarity matrix into a dissimilarity matrix via $\delta_{ij} = (s_{ii} + s_{jj} - 2s_{ij})^{1/2}$, and applying classical multidimensional scaling gives the configuration shown in Fig. 4.3. The horseshoe shape is very clear. The chronology can be read off as one proceeds around the horseshoe from one vertex to another. The reason that the 'graves' are not simply arranged linearly is because the dissimilarity between, for example, the first and the last grave, 1 and 51, is the same as that between other pairs of graves that are closer in time, such as 13 and 39. □

Kendall also gives a real example involving 59 graves at the La Tene cemetery at Münsingen-Rain using an incidence matrix for 70 varieties of anklets, bracelets, etc.

Assessing distortion in MDS configurations

4.3 *Residuals*, the differences between observed values and the corresponding values predicted by some model, are common in many areas of statistics, particularly regression, where they can be useful in indicating outliers, model deficiencies, etc. Residuals also have a role to play in applications of MDS, in highlighting aspects of the fit of the derived solution. The further a fitted distance value is from its corresponding disparity value, the worse the fit, and this will be reflected in the size of the residual, r_{ij} defined as

$$r_{ij} = d_{ij} - \hat{d}_{ij} \tag{4.2}$$

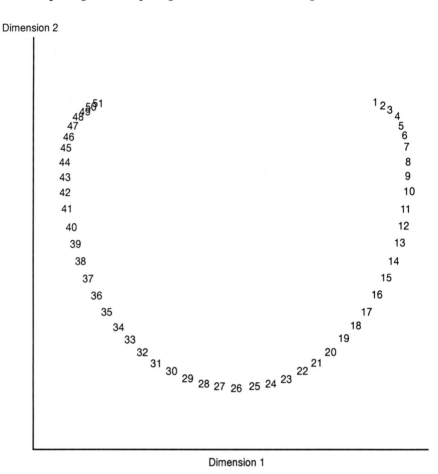

Figure 4.3   Scaling solution for Kendall's generated similarity matrix for 51 'graves'.

Presenting the residuals in the form of a lower triangular matrix, matching the observed proximity matrix, often aids in interpretation. Additionally, rounding, followed by multiplication by 10 (see Coxon, 1982b), can be helpful in clarifying the picture (More formal methods of modelling residuals in the context of MDS will be considered in Chapter 5, Section 5.5.)

A further procedure for assessing which parts of a scaling solution have the worst fit is to look at the contribution which each point makes to the overall stress value, i.e. for each point i, calculate S_i given by

$$S_i = \sum_{\substack{j=1 \\ j\neq i}}^{n} (d_{ij} - \hat{d}_{ij})^2 \tag{4.3}$$

Example 4.3

To illustrate the use of residuals in multidimensional scaling, those for the two-dimensional non-metric scaling solution for the World War II politicians data (see Table 3.6) are presented in Table 4.2. The largest residuals involve Tito and Franco, Chamberlain and Churchill, and Attlee and Eisenhower. Perhaps the explanation here is that these pairs involve one 'dominant' historical figure and one who made less impact on a world stage and the subject finds assigning dissimilarities to such pairs difficult.

A *dot plot* of the S_i values defined by (4.3) for the two-dimensional non-metric scaling of the World War II dissimilarities data is given in Fig. 4.4; Tito, Chamberlain and Franco have the largest values. Perhaps the subject finds judgements involving these three individuals particularly difficult since they are rather more 'minor' figures in the context of the Second World War than the remainder. □

In some cases the poor fit of particular stimuli might be explained by the optimization procedure encountering a local minimum, in the sense that some other configuration may exist with lower stress, which would locate the worst-offending points in another position but would be substantially similar in other respects. The only way to check whether this explanation is likely is to rerun the procedure with a different initial configuration; for example, the current solution with the poorly fitted points relocated to where the investigator thinks more appropriate.

A further useful device for highlighting possible distortions in an MDS solution is the *minimum spanning tree*, which can be defined thus.

Suppose *n* points are given (possibly in many dimensions), then a *tree* spanning these points, i.e. a *spanning tree*, is any set of straight line segments joining pairs of points such that:

(1) no closed loops occur;
(2) each point is visited by at least one line;
(3) the tree is connected, i.e. it has paths between any pairs of points.

Table 4.2 Residuals from non-metric scaling of dissimilarities of Second World War politicians

	1	2	3	4	5	6	7	8	9	10	11	12
1	0											
2	−9	0										
3	3	1	0									
4	6	3	−1	0								
5	−1	−2	3	2	0							
6	−3	−2	0	−10	4	0						
7	−7	−6	−2	2	−5	−3	0					
8	1	3	−8	−1	0	1	−3	0				
9	0	1	−2	−4	−1	9	0	−1	0			
10	0	2	4	−4	2	−1	−2	2	0	0		
11	2	4	−10	2	−1	−3	0	−7	1	8	0	
12	1	−1	5	−1	−4	−1	11	2	−3	−9	−3	0

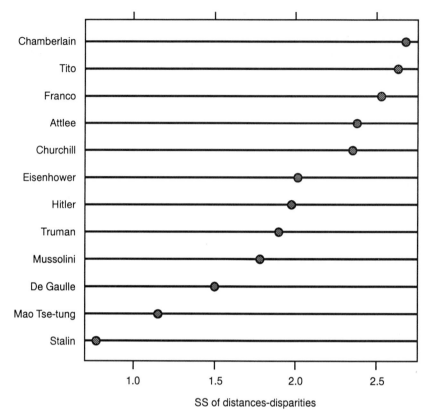

Figure 4.4   Dot plot of diagnostic terms for non-metric scaling solutions of World War II politicians.

The length of a tree is defined to be the sum of the lengths of its segments, and when a set of n points and the lengths of all $\binom{n}{2}$ segments are given, then the minimum spanning tree of the n points is defined as the spanning tree of minimum length. Algorithms to find the minimum spanning tree of a set of points given the distances between them are given in Prim (1957), Loberman and Weinberg (1957) and Gower and Ross (1969).

The links of the minimum spanning tree of the original dissimilarity matrix may be plotted on to the two-dimensional scaling representation in order to identify possible distortions produced by the scaling solution. Such distortions are indicated when nearby points on the plot are not joined by an edge of the tree.

Example 4.4

As an illustration of this approach, the minimum spanning tree of the dissimilarity matrix for populations of water voles in Table 3.7 is shown in Fig. 4.5(a), plotted onto the two-dimensional solution obtained from non-metric multidimensional scaling (see Chapter 3). The plot indicates, for example, that the

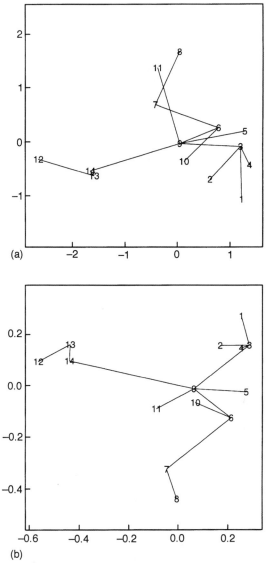

Figure 4.5 (a) Non-metric scaling for water voles data showing minimum spanning tree. (b) Classical scaling for water voles data showing minimum spanning tree.

apparent closeness of populations, Germany (9) and Norway (10) suggested by the points representing them in the MDS solution, does not reflect accurately their calculated dissimilarity; the links of the minimum spanning tree show that the Aberdeen (5) and Eilean Gamhna (6) populations are actually more similar to the German water voles than those from Norway. Again, the Pyrenees I (11) population is very poorly placed in the two-dimensional non-metric solution.

Figure 4.5(b) shows the two-dimensional solution obtained from classical scaling for these data, also with the minimum spanning tree imposed. There seems to be substantially less distortion with this solution; in particular the Pyrenees I population is more appropriately sited.   □

Spence and Lewandowsky (1989) illustrate that proximity judgements may be particularly susceptible to the problem of outlying observations (even a single outlier may dramatically distort a solution from classical scaling or metric scaling) and suggest a 'robust' approach to scaling that is not affected by such a problem.

Matching configurations

4.4   Since there are a large number of potential similarity and dissimilarity measures (see Chapter 2) and a variety of multidimensional scaling techniques (see Chapter 3), the possibility exists of generating many different geometrical configurations to represent the same set of stimuli. Two different configurations will be obtained, for example, if two different techniques are applied to the same proximity matrix, or if a single technique is applied to two proximity matrices for a set of stimuli, in which the elements are defined in different ways.

Given two solutions, it is natural to ask how similar they are. The usual approach to judging the degree of similarity of alternative scaling solutions for a set of proximity data is *Procrustes analysis* or its generalization to comparisons of more than two configurations, *generalized Procrustes analysis*.

Procrustes analysis

4.4.1   First, the question of the name. According to Kirk (1974), Procrustes, 'The Smasher', caught travellers on the borders of Athens and by stretching or pruning them using rack or sword, made them an exact fit to his lethal bed. (Procrustes eventually experienced the same fate as that of his guests at the hands of Theseus.)

Suppose the two configurations to be compared are represented as $\mathbf{X}$ and $\mathbf{Y}$. Assume that both matrices are $n \times q$, i.e. q coordinate values for each of n stimuli. (If the two configurations to be compared have different dimensionalities, q_1 and q_2, with $q_2 < q_1$, the matrices $\mathbf{X}$ and $\mathbf{Y}$ can be made to have the same number of columns, by simply adding columns of zeros to the one corresponding to q_2.) The two configurations are compared by a residual sum of squares, minimized by allowing the points defined by $\mathbf{Y}$ to be rotated, reflected and translated relative to $\mathbf{X}$ (scaling will be dealt with later). Specifically the fit criterion is R^2 given by

$$R^2 = \min_{\mathbf{A},\mathbf{b}} \sum_{i=1}^{n} (\mathbf{x}_i - \mathbf{A}'\mathbf{y}_i - \mathbf{b})'(\mathbf{x}_i - \mathbf{A}'\mathbf{y}_i - \mathbf{b}) \qquad (4.4)$$

where $\mathbf{A}$ is a $q \times q$ orthogonal matrix.

Mardia *et al.* (1979) show that if both **X** and **Y** are centred at the origin, so that $\bar{\mathbf{x}} = \bar{\mathbf{y}} = \mathbf{0}$, the solution of (4.4) for **A** and **b** is

$$\hat{\mathbf{b}} = \mathbf{0}, \quad \hat{\mathbf{A}} = \mathbf{VU'} \tag{4.5}$$

leading to a value of R^2 at the minimum of

$$R^2 = \mathrm{tr}(\mathbf{XX'}) + \mathrm{tr}(\mathbf{YY'}) - 2\,\mathrm{tr}\,\boldsymbol{\Gamma} \tag{4.6}$$

where **V**, **U** and $\boldsymbol{\Gamma}$ are the matrices obtained from the singular value decomposition of **Y'X**, $\boldsymbol{\Gamma}$ being the diagonal matrix of singular values. R^2 can take the value zero only when the $\mathbf{y}_i$ can be rotated to fit the $\mathbf{x}_i$ exactly.

Example 4.5

To illustrate the use of the Procrustes procedure consider the two-dimensional configurations obtained from applying metric and non-metric scaling to the World War II dissimilarities as reported in the previous chapter. The two

Table 4.3 Two-dimensional solutions from metric and non-metric scaling of dissimilarities of Second World War politicians

(a) Metric scaling solution

	Dimension 1	Dimension 2
Hitler	1.19	0.66
Mussolini	1.46	0.82
Churchill	0.02	−0.75
Eisenhower	−1.10	−1.07
Stalin	0.39	1.49
Attlee	0.65	−1.74
Franco	0.99	0.28
De Gaulle	0.02	−0.19
Mao Tse-tung	−1.52	1.40
Truman	−1.62	−0.17
Chamberlain	0.64	−0.90
Tito	−1.12	0.19

(b) Non-metric scaling solution

	Dimension 1	Dimension 2
Hitler	1.19	0.72
Mussolini	1.43	0.80
Churchill	0.17	−0.67
Eisenhower	−1.12	−1.10
Stalin	0.39	1.48
Attlee	0.63	−1.77
Franco	1.04	0.29
De Gaulle	0.05	−0.31
Mao Tse-tung	−1.54	1.35
Truman	−1.58	−0.16
Chamberlain	0.55	−0.85
Tito	−1.22	0.25

Table 4.4   Results of applying Procrustes procedure to scaling solutions in Table 4.3

$$\mathbf{\Gamma} = \begin{pmatrix} 13.01 & 0.00 \\ 0.00 & 10.95 \end{pmatrix}$$

$$\mathbf{U} = \begin{pmatrix} -0.99 & -0.13 \\ -0.13 & 0.99 \end{pmatrix}$$

$$\mathbf{V} = \begin{pmatrix} -0.99 & -0.12 \\ -0.12 & 0.99 \end{pmatrix}$$

$$\mathbf{A} = \begin{pmatrix} 1.00 & 0.01 \\ -0.01 & 1.00 \end{pmatrix}$$

solutions are reproduced here in Table 4.3. The results of the Procrustes analysis are given in Table 4.4. Here the value of $R^2 = 0.082$ is small as expected since we have already commented on the similarity of the two solutions in Chapter 3. The rotation matrix, $\mathbf{A}$ for this example (see Table 4.4) is almost the identity matrix. □

If the scales of the two configurations to be compared are different then the transformation to be considered should be

$$c\mathbf{A}'\mathbf{y}_i + \mathbf{b} \tag{4.7}$$

where $c > 0$. Mardia *et al.* (1979) show that c is estimated as

$$\hat{c} = (\mathrm{tr}\mathbf{\Gamma})/(\mathrm{tr}\mathbf{YY}') \tag{4.8}$$

and the estimates of $\mathbf{A}$ and $\mathbf{b}$ remain the same. This transformation is called the *Procrustes rotation with scaling* of $\mathbf{Y}$ relative to $\mathbf{X}$. The new minimum residual sum of squares is given by

$$R^2 = \mathrm{tr}(\mathbf{XX}') + \hat{c}^2 \, \mathrm{tr}(\mathbf{YY}') - 2\hat{c} \, \mathrm{tr}(\mathbf{\Gamma}) \tag{4.9}$$

This procedure is not symmetrical with respect to $\mathbf{X}$ and $\mathbf{Y}$. Symmetry can be obtained by selecting scaling so that

$$\mathrm{tr}(\mathbf{XX}') = \mathrm{tr}(\mathbf{YY}') \tag{4.10}$$

Example 4.6

As a further, more substantial, example of Procrustes analysis, the German road distances introduced in the previous chapter will be used. The latitudes and longitudes of each town are given in Table 4.5. Treating these quantities as planar coordinates $\mathbf{X}(27 \times 2)$ (assuming that the curvature of the Earth has little effect), and the two-dimensional solution obtained from classical MDS (see Chapter 3) as $\mathbf{Y}$, a Procrustes analysis was performed with the results shown in Table 4.6.

A non-metric scaling of the German road distances gives the two-dimensional configuration shown in Table 4.7. Application of the Procrustes procedure to this solution gives the results shown in Table 4.8. The resulting R^2 values for both classical and non-metric scaling are very similar.

Table 4.5 Latitudes and longitudes of German cities

	Latitude	Longitude
Aachen	50.77	6.10
Berlin	52.53	13.42
Bielefeld	52.03	8.53
Bonn	50.73	7.10
Bremen	53.08	8.80
Dortmund	51.53	7.45
Dresden	51.05	13.75
Düsseldorf	51.21	6.78
Essen	51.45	6.95
Frankfurt	50.10	8.68
Freiburg	48.00	7.87
Hamburg	53.55	10.00
Hannover	52.53	9.73
Karlsruhe	49.00	8.40
Kassel	51.30	9.50
Kiel	54.33	10.13
Köln	50.93	6.95
Leipzig	51.33	12.42
Mainz	50.00	8.27
Mannheim	49.50	8.47
München	48.13	11.58
Münster	51.97	7.62
Nürnberg	49.45	11.08
Saarbrücken	49.25	6.97
Stuttgart	48.78	9.95
Würzburg	49.80	9.95
Wuppertal	51.25	7.17

Figure 4.6 shows the plots of the German cities after Procrustes analysis of the three configurations; the two scaling solutions are clearly in relatively good agreement both with each other and with the 'true' map of the cities. □

Generalized Procrustes analysis

4.4.2 Suppose now that it is required to compare more than two configurations represented as $X_1, X_2, \ldots, X_m$. One possibility would be to apply the

Table 4.6 Results of Procrustes procedure for classical scaling solution of German distance matrix and latitudes and longitudes of cities

$$\Gamma = \begin{pmatrix} 10490.25 & 0.00 \\ 0.00 & 8718.67 \end{pmatrix}$$

$$U = \begin{pmatrix} 0.7048 & -0.7094 \\ 0.7094 & -0.7048 \end{pmatrix}$$

$$V = \begin{pmatrix} 0.6060 & -0.7955 \\ 0.7955 & -0.6060 \end{pmatrix}$$

$R^2 = 11.93, c = 0.0085$

Table 4.7 Two-dimensional non-metric scaling solution for German road distances

	Dimension 1	Dimension 2
Aachen	−0.02	1.28
Berlin	−1.31	−1.61
Bielefeld	−0.77	0.21
Bonn	0.20	0.93
Bremen	−1.58	0.10
Dortmund	−0.62	0.76
Dresden	−0.37	−1.95
Düsseldorf	−0.13	1.02
Essen	−0.38	0.86
Frankfurt	0.64	0.12
Freiburg	2.09	0.13
Hamburg	−1.86	−0.17
Hannover	−1.06	−0.19
Karslruhe	1.35	0.11
Kassel	−0.31	−0.25
Kiel	−2.33	−0.31
Köln	0.06	0.91
Leipzig	−0.53	−1.41
Mainz	0.69	0.24
Mannheim	1.04	0.16
München	1.72	−1.45
Münster	−0.91	0.84
Nürnberg	1.05	−1.05
Saarbrücken	1.33	0.79
Stuttgart	1.54	−0.43
Würzburg	0.76	−0.50
Wuppertal	−0.30	0.85

Stress = 0.04

Table 4.8 Results of Procrustes procedure for non-metric scaling solution of German distance matrix and latitudes and longitudes of cities

$$\Gamma = \begin{pmatrix} 51.18 & 0.00 \\ 0.00 & 42.68 \end{pmatrix}$$

$$U = \begin{pmatrix} 0.7057 & 0.7085 \\ 0.7085 & -0.7057 \end{pmatrix}$$

$$V = \begin{pmatrix} -0.7901 & -0.6130 \\ -0.6130 & 0.7901 \end{pmatrix}$$

$R^2 = 11.86, c = 1.74$

procedure described in the previous section to each pair of configurations and form the $m \times m$ symmetric matrix of pairwise goodness-of-fit sums of squares. This matrix might then be analysed by metric or non-metric scaling methods to uncover the relationships between the various configurations. (Gower, 1975, demonstrated that the matrix of sums of squares from the pairwise Procrustes

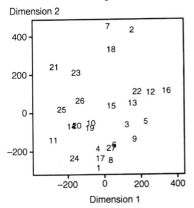

German cities longitude and latitude

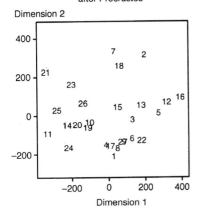

German cities, classical scaling after Procrustes

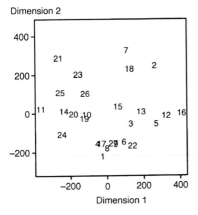

German cities, non-metric scaling after Procrustes

Figure 4.6 Map of German cities and classical and non-metric scaling solutions after Procrustes analysis.

analyses is a metric, but was unable to prove that it is Euclidean.) An example of this approach is given in Krzanowski and Marriot (1994).

An alternative form of analysis of the m configurations is a generalization of the Procrustes idea in which all m sets are simultaneously translated, rotated, and reflected so that a goodness-of-fit criterion is optimized. Here the fit criterion is defined as

$$R^2 = \text{tr} \sum_{i=1}^{m} (\mathbf{X}_i\mathbf{H}_i - \mathbf{Y})'(\mathbf{X}_i\mathbf{H}_i - \mathbf{Y}) \tag{4.11}$$

where $\mathbf{H}_1, \mathbf{H}_2, \ldots, \mathbf{H}_m$ are orthogonal matrices and

$$\mathbf{Y} = \frac{1}{m} \sum_{i=1}^{m} \mathbf{X}_i\mathbf{H}_i \tag{4.12}$$

Gower (1975) shows that $\mathbf{H}_i$ is given by

$$\mathbf{H}_i = \mathbf{V}_i\mathbf{U}_i \tag{4.13}$$

where the matrices $\mathbf{V}_i$ and $\mathbf{U}_i$ arise from the singular-value decomposition of $\mathbf{X}_i'\mathbf{Y}$, i.e.

$$\mathbf{X}_i'\mathbf{Y} = \mathbf{V}_i'\mathbf{\Gamma}\mathbf{U}_i \tag{4.14}$$

where $\mathbf{\Gamma}$ is a diagonal matrix.

Since the centroid matrix, $\mathbf{Y}$, is not known, this does not give an immediate method for calculating $\mathbf{H}_i$, but suggests the following simple iterative procedure.

Step 1. Set $\mathbf{Y} = \mathbf{X}_1$; for $i = 2, 3, \ldots, m$, rotate $\mathbf{X}_i$ to fit $\mathbf{Y}$ and re-evaluate $\mathbf{Y}$ as the mean of $\mathbf{X}_1, \ldots, \mathbf{X}_i$.

Step 2. Rotate $\mathbf{X}_i$, $i = 1, 2, \ldots, m$ to fit $\mathbf{Y}$; recompute $\mathbf{Y}$, evaluate R^2 and repeat this step until some convergence criterion is satisfied.

Step 3. Refer $\mathbf{Y}$ and each $\mathbf{X}_i$ to the principal axes of $\mathbf{Y}$, to give a unique average picture to be compared with each individual set of observations.

Examples of the applications of generalized Procrustes analysis are given in Gower (1975) and Krzanowski and Marriot (1994). Here, however, an account will be given of the application of an adaptation of the method known as *weighted generalized Procrustes analysis*, reported in Everitt and Gower (1981).

In the weighted form of generalized Procrustes, associated with each $\mathbf{X}_i$ matrix is a diagonal matrix of 'weights', $\mathbf{N}_i$, one weight for each of the n rows of $\mathbf{X}_i$. The aim now is to find orthogonal matrices $\mathbf{H}_i$ that minimize

$$R_W^2 = \text{tr} \sum_{i=1}^{m} (\mathbf{X}_i\mathbf{H}_i - \mathbf{Y})'\mathbf{N}_i(\mathbf{X}_i - \mathbf{H}_i\mathbf{Y}) \tag{4.15}$$

where $\mathbf{Y}$ is now given by

$$\mathbf{Y} = \left[\sum_{i=1}^{m} \mathbf{N}_i\right]^{-1} \sum_{i=1}^{m} \mathbf{N}_i\mathbf{X}_i\mathbf{H}_i \tag{4.16}$$

The iterative algorithm for generalized Procrustes may easily be adapted for the weighted version as follows.

Step 1. Find an initial value of $\mathbf{Y}$. (Here, using any particular $\mathbf{X}_i$ matrix may no longer be possible; see below.)

Step 2. Rotate $\mathbf{N}_i^{1/2}\mathbf{X}_i$, $i = 1, 2, \ldots, m$ to fit $\mathbf{N}_i^{1/2}\mathbf{Y}$; recompute $\mathbf{Y}$, evaluate R_W^2 and repeat until some convergence criterion is satisfied.

Step 3. Refer $\mathbf{Y}$ and each $\mathbf{X}_i$ (if required) to the principal axis of $\mathbf{Y}$.

Example 4.7

The application of weighted generalized Procrustes analysis described by Everitt and Gower (1981) involved data collected in an investigation concerned with the substitution of vision in blind people. One approach to the problem described by Brindley and Lewin (1968) is the *cortical visual prosthesis*. This device is an implanted electrical stimulator, with an array of electrodes laid on that surface of the brain concerned with vision, namely the visual cortex. Stimulation at one or more electrodes can be applied under control from the exterior via inductive ('radio') links across the skin. When a train of electrical pulses is delivered at one of the electrodes, the patient reports the presence of a spot or patch of light sensation, termed a *phosphene*. Stimulation through different electrodes results in phosphenes in different positions in the visual field, and the characteristic positions of different phosphenes are largely consistent from test to test. The notion then is to combine phosphenes into more elaborate visual sensations which could be of practical use to the patient, by stimulation of more than one electrode at once. Ordinary print, for example, might be read using a camera suitably arranged to control the pattern of stimulation at different combinations of electrodes.

Before beginning to use the prosthesis in this way, the position of each phosphene in the visual field has first to be determined, to give what is termed a *phosphene map*. Since the patient is blind, the positions of the phosphenes can only be observed relative to each other, so that the phosphene map must be built up from individual observations of the angle and separation between various different phosphene pairs. Such observations are conveniently made by stimulating two phosphenes in sequence and asking the subject to record the angle and distance of the centre of the second phosphene from the centre of the first, taking the first as if at the centre of a clock face; angles are therefore reported by the subject in 'clock minutes.' Distances are recorded in inches of separation at the apparent distance of the phosphenes from the eye. Observations of the same phosphene pair may vary somewhat from time to time and observations of phosphene pairs do not fit perfectly into a phosphene map. Consequently a procedure is needed to find the best fitting map for a set of observations. One approach is described in Everitt and Rushton (1978), but here we shall concentrate on the weighted generalized Procrustes procedure suggested by Everitt and Gower (1981).

Each observation of distance and angle may be converted to the corresponding Cartesian coordinates and these are averaged to give the coordinates of a set of phosphenes, $i_1, i_2, \ldots, i_{n_i}$ relative to phosphene i as the origin – see Table 4.9. These coordinates are stored in $\mathbf{X}_i$, so here $p = 2$, $m = n$, where n is the number

Table 4.9   Phosphene i – observations relative to phosphenes $i_i, i_2, \ldots, i_{n_i}$

Phosphene pair	Observations
$i - i_1$	$(d_{ii_1 1}, a_{ii_1 1}) \ldots (d_{ii_1 n_{ii_1}}, a_{ii_1 n_{ii_1}})$
$i - i_2$:	$(d_{ii_2 1}, a_{ii_2 1}) \ldots (d_{ii_2 n_{ii_2}}, a_{ii_2 n_{ii_2}})$
$\vdots$	
$i - i_{n_i}$:	$(d_{ii_{n_i} 1}, a_{ii_{n_i} 1}) \ldots (d_{ii_{n_i} n_{iin_i}}, a_{ii_{n_i} n_{iin_i}})$

d represents the observed distance and a the observed angle.

of phosphenes. (Some rows of $\mathbf{X}_i$ will contain no entries since phosphene i is not observed relative to all other phosphenes.)

The weight matrices, $\mathbf{N}_i$, arise because of the differing number of observations made on each phosphene pair, and contain the number of times phosphene j is observed relative to phosphene i. Some values may be zero.

To begin the iterative algorithm for weighted Procrustes, we need an initial value of $\mathbf{Y}$ which contains an entry for each row, i.e. we need initial coordinate values for all phosphenes. None of the $\mathbf{X}_i$ matrices satisfied this requirement, but

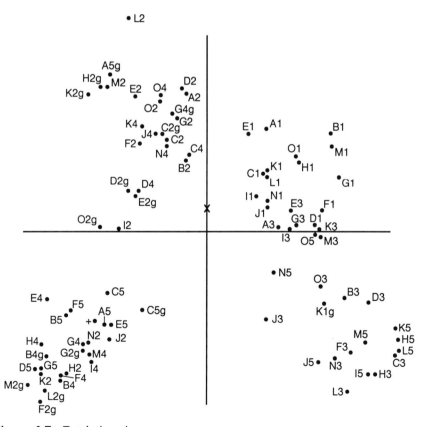

Figure 4.7   Total phosphene map.

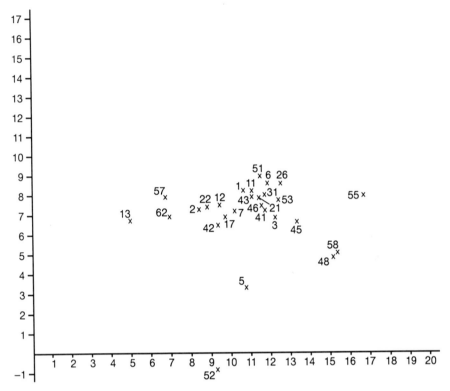

Figure 4.8 Individual phosphene map.

the solution obtained from the method described in Everitt and Rushton (1978) did, and was therefore used to provide the starting configuration for weighted Procrustes.

The weighted Procrustes algorithm took some 28 iterations to converge. The final phosphene map is shown in Fig. 4.7 and two of the final 'individual' phosphene maps in Figs 4.8 and 4.9. □

Krzanowski and Marriot (1994) draw attention to the fact that the various Procrustes procedures discussed above are purely descriptive in nature and what is lacking is any guideline as to the amount of sampling variability to be expected in the statistic R^2, which would allow more objective statements to be made about the similarity or otherwise of two matched configurations.

Procrustes methods have become of great importance in the statistical analysis of shape, where the shape of a geometrical figure is commonly understood to refer to those geometrical attributes that remain unchanged when the figure is translated, rotated and scaled. In many instances the geometrical figure is described by N labelled points in K dimensions, the labelled points often being referred to as *landmarks* (Bookstein, 1978). Such data arise in biology, archaeology, astronomy, geology etc, and suitable techniques for analysis are described in Bookstein (1991), Goodall (1991) and Commandeur (1991).

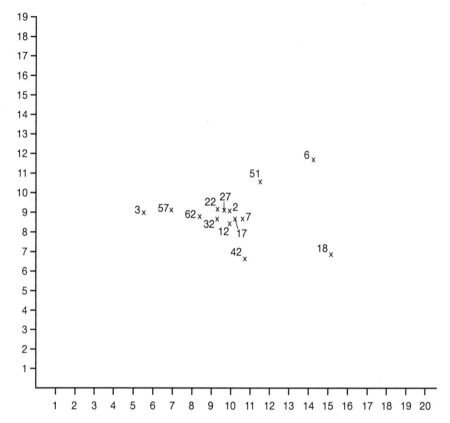

Figure 4.9   Individual phosphene map.

Summary

4.5   Interpreting, diagnosing and comparing scaling solutions are important aspects of using MDS in practice. The examination of residuals and the use of the minimum spanning tree can be very helpful in highlighting distortions in derived configurations. Procrustes analysis can be used to compare competing solutions and recently has also become of great importance in the more general problem of comparing 'shape'. Procrustes procedures are also occasionally used as the basis for *individual differences scaling* to be described in Chapter 6.

An alternative to Procrustes analysis for comparing scaling solutions, which has been suggested by a number of authors, is *constrained* multidimensional scaling in which some of the points or some of the dimensions are constrained to equal those corresponding to some hypothesized structure. The stress of the resulting configuration can then be used to assess if the data appear to be consistent with the hypothesis. For details, readers are referred to Lee and Bentler (1980), Lee (1984), Borg and Lingoes (1980) and Cox and Cox (1994).

5

Three-way Multidimensional Scaling

Introduction

5.1 In Chapter 3 some scaling methods applicable to a single matrix of proximities were discussed. However, in many situations encountered in practice, more than one proximity matrix will be available. Examples include proximity matrices for different subjects, different occasions or different conditions. A simple example is shown in Table 5.1.

Here, the dissimilarities for four drinks as judged by three subjects are given. The three matrices could be averaged to give a single proximity matrix for analysis by one of the methods of Chapter 3, but such an approach assumes that differences between the subjects' dissimilarities are simply due to error and that there are no systematic differences between the three observed matrices. Another similar approach is to use the *replicated multidimensional scaling* method suggested by McGee (1968), which minimizes a combined stress value over the matrices. It is, however, very clear from direct inspection of Table 5.1 that in this example there *are* systematic differences between subjects. Subject 2, for example, considers the colour to be more important than the type of drink (wine or juice) when judging dissimilarity. For subjects 1 and 3, type appears to be more important than colour. Subject 1 in particular considers the ingredients of the drink to be more important than any other characteristic, as reflected by his judgement that white wine is more like red grape juice than like apple juice.

Table 5.1 Dissimilarity matrices for four drinks (rw = red wine, ww = white wine, rgj = red grape juice, aj = apple juice) and three subjects

	Subject 1				Subject 2				Subject 3			
	rw	ww	rgj	aj	rw	ww	rgj	aj	rw	ww	rgj	aj
rw	0				0				0			
ww	2	0			3	0			1	0		
rgj	1	3	0		1	6	0		3	6	0	
aj	6	5	4	0	5	2	4	0	5	4	2	0

In this example, ignoring individual differences by analysing an average dissimilarity matrix or using replicated multidimensional scaling would clearly be misleading. In most studies involving multiple proximity matrices, it is precisely the individual differences that are of most interest, and so what is ideally required is some appropriate summary of the way in which individual subjects differ from each other. With this in mind, most methods designed specifically for three way data, therefore, operate by deriving an overall *group stimulus space* together with individual *private spaces* or *personal spaces* for each subject. The latter are allowed to differ from the former by a specified class of transformations, the parameters of which provide the required summary of the individual differences. These models, also known as *individual differences* models, differ in the extent to which private spaces are allowed to differ from the group stimulus space.

We will assume there are n stimuli and m 'subjects' (which will be a convenient term for the source of the individual matrices, also referred to as the *third way*, whether this is individuals, conditions, occasions etc.). Table 5.1 illustrates the general form of three-way data. The dissimilarities for subject s will be denoted by $\delta_{ij}^{(s)}$. The solutions from an individual differences model will generally consist of two parts, namely the $n \times q$ group stimulus configuration $\mathbf{X}$ and a set of parameters required to generate the separate subject configurations $\mathbf{X}^{(1)}$, $\mathbf{X}^{(2)}, \ldots, \mathbf{X}^{(m)}$. This chapter describes individual differences models in order of increasing flexibility in the allowed transformations from the stimulus space to the private spaces.

Weighted Euclidean model

5.2   Although there was earlier work on the problem (see, for example Tucker and Messick, 1963; Ross, 1966), the first practically successful individual differences scaling model, the *weighted Euclidean model*, was suggested independently by several authors in the late 1960s, early 1970s (Carroll and Chang, 1970; Horan, 1969; Bloxom, 1968).

The model assumes that there exists a group stimulus space $\mathbf{X}$ in which the stimuli are uniquely representable by a set of points; this implies that each subject is assumed to identify the same sources of variation among the stimuli. The subjects may, however, differ in the relative importance they attach to these sources and this feature is captured in the model by allowing each subject to have a unique set of weights or *saliences* to apply to each of the axes in the group stimulus space. These weights, $w_k^{(s)}$, $k = 1, \ldots, q$, $s = 1, \ldots, m$ summarize the individual differences between subjects and form the *subject space* or *weight space*. Each subject's private space $\mathbf{X}^{(s)}$ is found from $\mathbf{X}$ simply by multiplying the group stimulus coordinates, x_{ik}, by the square roots of the corresponding weights, i.e.

$$x_{ik}^{(s)} = \sqrt{w_k^{(s)}} x_{ik} \qquad (5.1)$$

The Euclidean distance between stimuli i and j for subject s is therefore given by

$$d_{ij}^{(s)} = \left\{ \sum_{k=1}^{q} w_k^{(s)} (x_{ik} - x_{jk})^2 \right\}^{1/2} \qquad (5.2)$$

or equivalently by

$$(d_{ij}^{(s)})^2 = (\mathbf{x}_i - \mathbf{x}_j)'\mathbf{W}_s(\mathbf{x}_i - \mathbf{x}_j) \qquad (5.3)$$

where $\mathbf{W}_s$ is a $(q \times q)$ diagonal matrix of the weights of subject s. There are a number of ways that the model can be fitted. The original method consisted of an adaptation of classical scaling where the inner products of the derived configuration are fitted to the double-centred dissimilarity matrix. Later developments involved methods more similar to metric and non-metric scaling, as described in Chapter 3, in which the distances of the derived configuration are fitted to transformed dissimilarities (disparities), i.e. some measure of discrepancy between distances and disparities is minimized. Details of the inner products approach will be described in Section 5.2.1 and the distance method is discussed in Section 5.2.3.

Deriving the group stimulus and weight matrices; the inner products approach

5.2.1   The first step in fitting the weighted Euclidean model is to convert the observed dissimilarity matrices to inner product matrices $\mathbf{B}_s = [b_{ij}^{(s)}]$ by double centring (cf. classical scaling, Chapter 3). It follows from the definition of the weighted Euclidean model in (5.1) that the corresponding inner products predicted by the model are given by $\sum_k x_{ik} w_k^{(s)} x_{jk}$. We have therefore

$$b_{ij}^{(s)} = \sum_k x_{ik} w_k^{(s)} x_{jk} + \epsilon_{ij}^{(s)} \qquad (5.4)$$

where ϵ_{ij}^s is the error term. The objective is to minimize the sum of squared residuals

$$\sum_{i,j,s} \left(b_{ij}^{(s)} - \sum_k x_{ik} w_k^{(s)} x_{jk} \right)^2 \qquad (5.5)$$

between the observed and predicted inner products.

In order to ensure that all subjects are weighted equally in the minimization, the matrices $\mathbf{B}_s$ are first normalized to have sum of squares equal to one

$$\sum_{ij} (b_{ij}^{(s)})^2 = 1 \qquad (5.6)$$

In matrix notation, (5.4) may be written as

$$\mathbf{B}_s = \mathbf{X} \mathbf{W}_s \mathbf{X}' + \mathbf{E}_s, \quad s = 1, \ldots, m \qquad (5.7)$$

The solution is found by treating the stimulus matrix $\mathbf{X}$ and its transpose $\mathbf{X}'$ in (5.7) as if they were distinct. Let $\mathbf{X}^L$ and $\mathbf{X}^R$ denote the matrices on the left and right respectively. Starting with an initial configuration, the algorithm iteratively updates each of the matrices $\mathbf{W}$, $\mathbf{X}^L$ and $\mathbf{X}^R$ in turn by minimizing (5.5), keeping the other two matrices equal to their current estimates. Each of the updates is simply a linear least squares problem. We briefly show this for the estimation of $\mathbf{W}$ (see Carroll and Chang, 1970, for further details of the algorithm).

In order to estimate $\mathbf{W}$, define a new matrix of inner products $\mathbf{B}^*$ having n^2 rows, one for each pair of stimuli and m columns, one for each subject, and let $\mathbf{E}$ ($n^2 \times m$) be the corresponding matrix of residuals. Also, let $\mathbf{G}$ be a matrix with n^2 rows and q columns given by $G_{lk} = x_{ik}x_{jk}$ and $\mathbf{W}^*$ ($q \times m$) a matrix of weights, $W_{ks}^* = w_k^{(s)}$.

Then (5.7) can be rewritten as

$$\mathbf{B}^* = \mathbf{G}^*\mathbf{W}^* + \mathbf{E} \tag{5.8}$$

and the least squares solution is simply

$$\mathbf{W}^* = (\mathbf{G}^{*\prime}\mathbf{G}^*)^{-1}\mathbf{G}^{*\prime}\mathbf{B}^* \tag{5.9}$$

The weights obviously should not be negative because this would result in negative squared distances in (5.3). Since, however, the algorithm described can lead to negative weights, such occurrences are dealt with by arbitrarily setting negative weights to zero. (Berge *et al.*, 1993, suggest a computational solution to the problem of negative weights.)

We now briefly show that the squared length of each subject's weight vector, $\sum_k (w_{ij}^{(s)})^2$ is approximately equal to the proportion of variation of the subject's observed inner products explained by the weighted Euclidean model.

For simplicity, consider one row of (5.8) in two dimensions and for one subject

$$b_{ij}^{(s)} = ((x_{i1}x_{j1}), (x_{i2}x_{j2})) \begin{pmatrix} w_1^{(s)} \\ w_2^{(s)} \end{pmatrix} + \epsilon_{ij}^{(s)} \tag{5.10}$$

The usual convention is to place the stimulus space origin at the mean

$$\sum_{i=1}^{n} x_{ik} = 0 \tag{5.11}$$

so that the means of the columns of the design matrix $\mathbf{G}^*$ are zero. Also, the axes are scaled so that the sum of squares of the coordinates is one for each dimension

$$\sum_{i=1}^{n} x_{ik}^2 = 1 \tag{5.12}$$

Clearly, this rescaling of the axes (multiplying the x_{ik} by constants c_k) is possible by making the corresponding adjustment to the weights (dividing the $w_k^{(s)}$ by c_k^2). The sum of squares about the mean due to the regression in (5.10) is therefore given by

$$\sum_{ij} (w_1^2(x_{i1}x_{j1})^2 + w_2^2(x_{i2}x_{j2})^2 + 2w_1w_2x_{i1}x_{i2}x_{j1}x_{j2})$$

$$= w_1^2 + w_2^2 + 2w_1w_2 \sum_i x_{i1}x_{i2} \sum_j x_{j1}x_{j2} \tag{5.13}$$

Since from (5.6), the total sum of squared scalar products is equal to one, this expression represents the proportion of variation in the scalar products accounted for by the regression. If the columns of $\mathbf{X}$ are columnwise orthogonal, the last term vanishes and the proportion of variation explained by the model is simply given by the squared length of the weight vector, $w_1^2 + w_2^2$.

Interpretation of group stimulus and weight spaces

5.2.2   A special property of the weighted Euclidean model is that the orientation of the dimensions of the stimulus space is unique (apart from permutations and reflections). The reason for this is that stretching and shrinking of the stimulus space to produce the private spaces is only possible in the directions of the dimensions. If the model is appropriate, i.e. the subjects do use the same criteria when judging similarities and there is substantial variation between subjects in the relative importance of these criteria, then the dimensions of the stimulus space should be directly interpretable as these criteria.

As we have already seen, the scaling of each of the dimensions is not unique however and is usually determined to achieve an equal spread of points along all dimensions (see (5.12)). This means that the relative scaling of the group stimulus axes does not in any way represent that of the average subject. A scaling that is representative of the average subject could be achieved by rescaling the weights as

$$w_k^{(s)*} = w_k^{(s)} \Big/ \sum_s w_k^{(s)}$$

When the stimulus space is rescaled accordingly, it represents the average subject.

In weight space, each subject may be represented as a vector pointing from the origin to that subject's set of weights. As we have already shown in Section 5.2.1, the squared lengths of the weight vectors are approximately equal to a measure of goodness of fit of the model to each subject's configuration. In the inner products method, the squared weight vectors are approximately equal to the proportion of variation in the inner products accounted for by the model and, as we will show in Section 5.2.3, in the distance method, the squared lengths of the weight vectors are exactly equal to the proportion of variation in the disparities explained by the model. The greater the angle between the weight vectors of the two subjects, the greater the difference in the relative importance the subjects assign to the dimensions. However, the distances between the vectors' endpoints have no straightforward interpretation.

Borg and Lingoes (1978) point out that it may be more useful to transform the weight space by taking the square roots of the weights. The distance between the endpoints of the vectors in this transformed weight space would then be equal to the distance between their private spaces defined by the square root of the sum of squared distances between corresponding points. Furthermore, the cosine of the angle between the weight vectors of two individuals would represent the correlation between the corresponding private spaces.

Example 5.1

As an example of the inner products method, it will now be applied to similarity data on 14 (now old-fashioned) computer languages for ten subjects. The data

Table 5.2   Similarities of 14 computer languages for subjects 8 (lower triangle) and 10 (upper triangle)

	alg	apl	ass	bas	c	cob	for	jcl	lis	mac	mat	pas	pli	sno
algol		3	6	5	9	4	6	0	4	0	3	9	9	4
apl	3		0	0	0	0	0	0	6	0	8	0	0	0
assembly	1	1		5	8	2	6	0	3	9	0	7	7	0
basic	4	3	1		5	4	9	0	2	2	5	5	5	4
c	5	4	2	2		4	6	0	0	8	4	8	8	4
cobol	3	2	1	1	3		4	0	5	2	2	5	6	7
fortran	5	4	1	5	4	4		0	2	5	6	8	7	2
jcl	2	1	6	3	1	2	1		0	0	0	0	0	0
lisp	3	2	0	1	2	1	1	1		3	6	3	3	6
machine	0	0	1	0	0	0	0	0	0		3	7	7	3
matrix	5	6	1	2	4	2	5	0	1	0		2	2	7
pascal	5	5	2	2	6	1	5	1	0	0	4		9	5
pli	6	4	2	2	6	4	5	1	0	0	4	6		5
snobol	2	1	0	1	1	1	1	0	2	0	2	1	1	

for subjects 8 and 10 are given in Table 5.2, and the fitted group stimulus space and subject space are given in Fig. 5.1.

The group stimulus space shows the overall structure which the private spaces share in common. Here, the dimensions may be interpreted as 'high level language' to 'low level language' (dimension 1) and 'popularity' or 'familiarity' (dimension 2). As expected, the dimensions are directly interpretable and no rotation of the space is required. The correlation between the two dimensions is only 0.07 so that the squared lengths of the weight vectors can be safely interpreted as representing the percentage of variation in the subject's scalar products explained by the model. This goodness of fit measure does not vary greatly between subjects; the mean correlation between observed and fitted scalar products is 0.68. The angle between the weight vectors of subjects 8 and 10 is quite large, indicating that these two subjects weight the dimensions quite differently.

The private spaces for these two subjects are shown in Fig. 5.2. These were obtained by scaling horizontal and vertical distances by the square roots of the appropriate weights. It is clear that subject 8 considers the level of the language more important than the popularity whereas the reverse is the case for subject 10. The correlations between the observed and fitted scalar products were 0.64 for subject 8 and 0.69 for subject 10.   □

Deriving the group stimulus and weight matrices; fitting squared distances to disparities

5.2.3   The inner products method is only appropriate for metric data with no missing values. Young, Takane, de Leeuw and Lewyckyi (Takane *et al.*, 1977; Young and Lewyckyi, 1979) have developed a more flexible approach which can be used to analyse non-metric data with missing values.

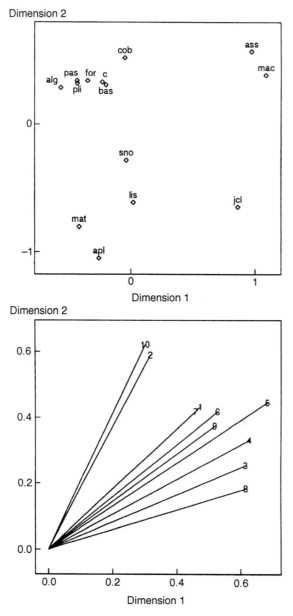

Figure 5.1 Groups stimulus space and subject space for computer languages.

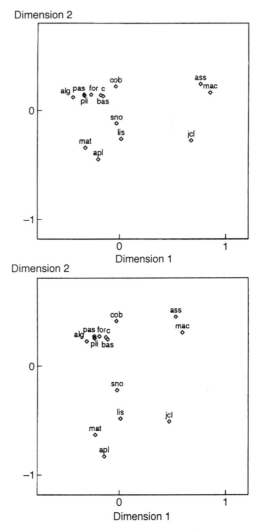

Figure 5.2   Private spaces for dissimilarities of computer languages for subjects 8 and 10.

This method aims to minimize a stress function analogous to the stress defined in Chapter 3 but with *squared* distances and disparities replacing the distances and disparities in the definition, to simplify the algorithm. This altered version of stress is known as *S-Stress*. For the weighted Euclidean model this is given by

$$\text{S-Stress} = \left(\frac{1}{m} \sum_s \frac{\sum_{ij} [(\hat{d}_{ij}^{(s)})^2 - (d_{ij}^{(s)})^2]^2}{\sum_{ij} (\hat{d}_{ij}^{(s)})^4} \right)^{1/2} \tag{5.14}$$

The flexibility of the method is due to the fact that, as in the metric and non-metric methods described in Chapter 3, the disparities $\hat{d}_{ij}^{(s)}$ are derived in a separate step using a transformation

$$f^{(s)}(\delta_{ij}^{(s)}) = \hat{d}_{ij}^{(s)} \qquad (5.15)$$

where $f^{(s)}$ might represent a metric (e.g. linear) or a non-metric, monotonic transformation. A non-metric analysis through monotonic regression is therefore a straightforward generalization of the metric method. Another advantage of minimizing S-Stress is that this is possible even when there are missing values, simply by restricting the summation in (5.14) for each subject to only those pairs of stimuli for which that subject has judged the dissimilarity.

We can also choose to allow the functional relationship $f^{(s)}$ between dissimilarities and disparities to differ between matrices, or to be the same for all matrices. The reason we might want to use different relationships is that each subject may be using his or her own subjective scale or 'response set', so that the dissimilarities in different matrices are not directly comparable. This property of the data is known as *matrix conditionality*. (As we shall see later, the analysis of row conditional data also becomes possible by using different transformations for different rows.) By contrast, MacCallum (1977) points out that the inner products method always implicitly treats the data as matrix conditional by normalizing the scalar products separately for each matrix (see (5.6)).

Starting with an initial configuration and a set of weights, the algorithm for minimizing S-Stress, called *alternating least squares*, iterates through the following steps until convergence. In the optimal scaling phase, the functions $f^{(s)}$ for the disparities $\hat{d}_{ij}^{(s)}$ are estimated by least squares for fixed stimulus configuration and weights. (Here, linear regression or monotone regression may be used depending on the measurement level of the data.) In the model estimation phase, $\mathbf{W}$ is first estimated by linear least squares for fixed $\mathbf{X}$ and then the coordinates of $\mathbf{X}$ are individually estimated by Newton–Raphson keeping $\mathbf{W}$ fixed. During the estimation of $\mathbf{W}$, the weights can be constrained to be non-negative. More details of the algorithm are given in Takane *et al.* (1977), but it is important to note that it uses the following normalization conventions

$$\sum_i x_{ik} = 0$$

$$\sum_i x_{ik}^2 = n$$

and

$$\sum_k (w_k^{(s)})^2 = \mathrm{cor}(\hat{d}_{ij}^{(s)}, d_{ij}^{(s)})^2$$

$$= r_s^2 \qquad (5.16)$$

Equation (5.16) is similar to (5.13) for the inner products approach in that it defines the goodness-of-fit measure represented by the squared length of the weight vectors. Here, the goodness-of-fit measure is the proportion of variance in the disparities explained by the distances. The equality in (5.16) is achieved (regardless of any correlations between dimensions) by multiplying each subject's disparities and weights by an appropriate contstant a_s, resulting in the distances in the private spaces being multiplied by that constant. (This rescaling does not affect the value of S-Stress or the correlation between distances and disparities.)

One criticism of the squared distance method is that it puts too much weight on large dissimilarities compared with small dissimilarities, although it might be argued that large dissimilarities are subject to larger errors than small ones (Ramsay, 1977). Weinberg and Menil (1993) compared the ability of the inner products and squared distances approaches to recover structure in simulated ordinal data (monotonically transformed distances for configurations with different numbers of points and subjects, corrupted by different amounts of error) and concluded that the inner products method is better, in most cases, at recovering the structure of the model even when compared with the non-metric version of the squared distance method.

Example 5.2

When attempting to fit the weighted Euclidean model to the drinks data in Table 5.1 using the inner products method, some of the estimated weights were negative. We therefore fitted the non-metric weighted Euclidean model to the data using the squared distance method. The resulting stimulus space is displayed in Fig. 5.3.

As expected, the dimensions can be directly interpreted as the type of drink and colour without having to rotate the figure first. Three measures of fit are given, S-Stress as defined in (5.14), r^2, the mean square correlation

$$r^2 = \frac{1}{m} \sum_s r_s^2$$

where r_s^2 is defined in (5.16), and Stress defined as

$$\text{Stress} = \left(\frac{1}{m} \sum_s \frac{\sum_{ij} [\hat{d}_{ij}^{(s)} - d_{ij}^{(s)}]^2}{\sum_{ij} (\hat{d}_{ij}^{(s)})^2} \right)^{1/2} \tag{5.17}$$

The weight space shows that subject 1 weights both dimensions equally, whereas subject 3 weights dimension 2 more strongly than dimension 1 and the reverse is the case for subject 2. Figure 5.4 shows the private spaces of the three subjects.

Note the large number of ties in the disparities for each subject. This is a typical occurrence in monotonic regression (see Chapter 3, Example 3.6) and can result in the ranking of the fitted distances within blocks of tied disparities being quite different to that of the observed disparities (compare columns two and four in the tables in Fig. 5.4).   □

Example 5.3

As a more substantive example of the weighted Euclidean model, we now look at a real dataset from psychology. The data were collected by Jacobowitz (see Young, 1974) to study the way language develops as children grow up. Subjects from four age groups judged the similarities of 15 body parts as follows. Each body part was selected in turn as a standard and the remaining body parts were

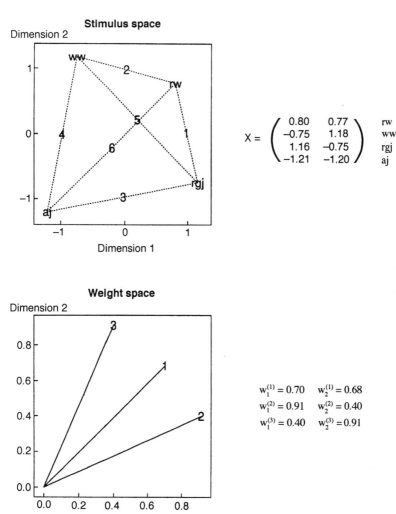

$$X = \begin{pmatrix} 0.80 & 0.77 \\ -0.75 & 1.18 \\ 1.16 & -0.75 \\ -1.21 & -1.20 \end{pmatrix} \begin{matrix} \text{rw} \\ \text{ww} \\ \text{rgj} \\ \text{aj} \end{matrix}$$

$w_1^{(1)} = 0.70 \quad w_2^{(1)} = 0.68$

$w_1^{(2)} = 0.91 \quad w_2^{(2)} = 0.40$

$w_1^{(3)} = 0.40 \quad w_2^{(3)} = 0.91$

S-Stress = 0.060, St ess = 0.029, r^2 = 0.98

Figure 5.3 Weighted Euclidean distance solution for drinks data.

ranked according to their similarity with the standard. Each subject therefore contributed a matrix whose *i*th row represents ranked similarities with the *i*th stimulus, the *i*th element being equal to zero. (Young calls these kinds of proximities *conditional rank order data*.) Here we use the data from 15 adults and 15 eight-year olds. The data for one of the children are shown in Table 5.3. (The full dataset is given in Young and Lewyckyi 1979.) The similarity matrices are asymmetric because they are row conditional. *Row conditionality* means that the values in each row are meaningful only relative to other values in the same row, not in other rows. For example, of all body parts, the one judged to be

Subject 1

(i,j)	$d_{ij}^{(s)}$	$\hat{d}_{ij}^{(s)}$	$\delta_{ij}^{(s)}$
1	1.29	1.30	1
2	1.34	1.34	2
4	2.01	2.10	5
3	2.02	2.10	4
5	2.26	2.10	3
6	2.35	2.35	6

S-Stress = 0.088 Stress = 0.043 $r_1^2 = 0.961$

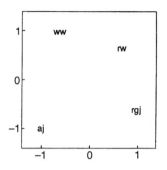

Subject 2

(i,j)	$d_{ij}^{(s)}$	$\hat{d}_{ij}^{(s)}$	$\delta_{ij}^{(s)}$
1	1.02	1.02	1
2	1.50	1.54	3
4	1.57	1.54	2
5	2.19	2.25	6
3	2.28	2.25	4
6	2.29	2.25	5

S-Stress = 0.038 Stress = 0.019 $r_2^2 = 0.995$

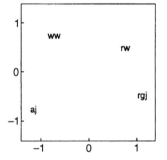

Subject 3

(i,j)	$d_{ij}^{(s)}$	$\hat{d}_{ij}^{(s)}$	$\delta_{ij}^{(s)}$
2	1.05	1.05	1
1	1.47	1.52	3
3	1.56	1.52	2
5	2.20	2.26	6
6	2.28	2.26	5
4	2.30	2.26	4

S-Stress = 0.038 Stress = 0.020 $r_3^2 = 0.994$

Figure 5.4 Plots of the personal spaces for drinks data together with the distances, disparities and dissimilarities, given in increasing order of the distances. The stimulus pairs (i,j) are coded as 1 = (rw, rgj), 2 = (rw, ww), 3 = (rgj, aj), 4 = (ww, aj), 5 = (ww, rgj), 6 = (rw, aj); (in order of increasing distances in stimulus space, see Fig. 5.3).

Table 5.3 The dissimilarity matrix of one of the 8-year old children for the 15 body parts

	cheek	face	mouth	head	ear	body	arm	elbow	hand	palm	finger	leg	knee	foot	toe
cheek	0	2	1	3	4	10	5	9	6	7	8	11	12	13	14
face	2	0	12	1	13	3	8	10	11	9	7	4	5	6	14
mouth	3	2	0	1	4	9	5	11	6	7	8	10	13	12	14
head	2	1	3	0	4	9	5	6	11	7	8	10	12	13	14
ear	10	1	11	2	0	6	3	4	5	12	13	7	8	14	9
body	14	12	9	6	13	0	8	7	5	10	11	1	4	2	3
arm	12	14	11	10	13	5	0	4	1	3	2	6	9	7	8
elbow	5	7	14	8	6	9	1	0	2	3	4	10	11	12	13
hand	13	11	12	10	14	9	3	4	0	1	2	6	5	7	8
palm	8	6	7	9	4	5	3	10	1	0	2	12	11	13	14
finger	14	5	13	6	9	12	3	4	1	2	0	7	8	10	11
leg	14	12	13	11	9	7	4	6	5	3	10	0	8	1	2
knee	12	11	14	10	13	4	5	8	6	7	9	1	0	2	3
foot	12	14	10	13	11	9	4	5	8	6	7	2	3	0	1
toe	13	8	9	11	14	3	6	5	7	10	12	2	4	1	0

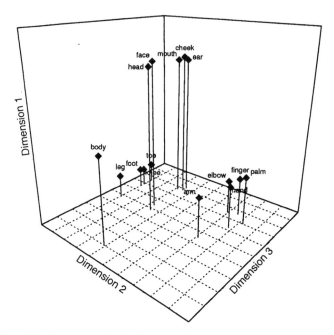

Figure 5.5   Stimulus space for body parts data.

most similar to 'body' is 'leg'. However, this does not imply that 'body' will be judged to be the most similar of all body parts to 'leg'. In fact, 'body' is only the seventh most similar body part to 'leg'. Another feature of this dataset is that it is ordinal since the subjects were asked simply to rank the body parts. This dataset has been analysed previously by Young (1974, 1994) who has kindly given us permission to repeat his analysis here.

We estimated the non-metric, weighted Euclidean model for row conditional data using the squared distance algorithm. The S-Stress values for the 2, 3, 4 and 5-dimensional solutions were 0.33, 0.27, 0.23 and 0.20 respectively. The group stimulus space for the three-dimensional solution is presented in Fig. 5.5.

Four distinct clusters are apparent, body, leg parts, head parts and arm parts. As Young points out, the solution has a hierarchical structure which is here represented by dimension 3. This hierarchical structure can be described by 'has a'. For example, moving back in the direction of dimension 3, we could say the body 'has an' arm which 'has a' hand which 'has a' finger. In dimension 3, the body is separated from the layer containing the attachments leg, arm and head, which is separated from the layers containing further attachments. Dimension 1 further distinguishes between the head parts and the limb parts and dimension 2 separates the limb parts into arm parts and leg parts. This is more clearly seen in the two-dimensional plots shown in Figs 5.6 and 5.7.

The weights of children and adults fall into different parts of the weight space shown in Fig. 5.8, implying that children and adults have different perceptions of the body parts. For example, compared with the children, the adults give very little weight to dimension 2. Figure 5.6 can therefore be thought of representing

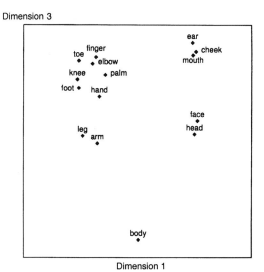

Figure 5.6 Dimensions 1 and 3 of the stimulus space (an exaggerated adult).

an exaggerated adult. Here, the hierarchical structure is important, but the distinction between arm and leg parts is not. In fact, corresponding arm and leg parts, for example knee and elbow and toe and finger, are close together. The children tend to have very low weights on dimension 3 so that Fig. 5.7 can be thought of as an exaggerated child. Here, the body is simply divided into the clusters body, head parts, arm parts and leg parts. ☐

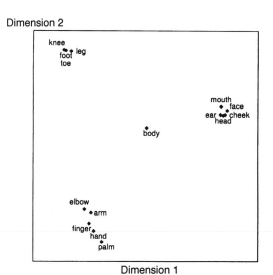

Figure 5.7 Dimensions 1 and 2 of the stimulus space (an exaggerated child).

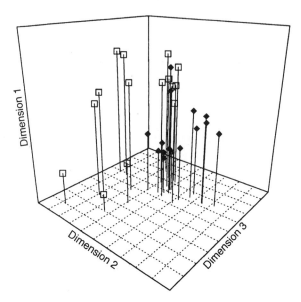

Figure 5.8   Weight space for body parts data. Squares: children; diamonds: adults.

Rotated and weighted Euclidean model

5.3   In the weighted Euclidean model, the weight matrix $\mathbf{W}_s$ in equation (5.3) is constrained to be diagonal. Carroll and Chang (1972) relaxed this constraint by allowing $\mathbf{W}_s$ to be any full rank, symmetric positive semi-definite matrix. The advantage of this more general model is that it allows the group stimulus configuration to be deformed by differentially scaling different sets of orthogonal 'axes' for different subjects whereas, in the Euclidean model, stretching and shrinking of the space was only possible along one set of axes, the coordinate axes of the stimulus space. The directions of the stimulus space dimensions are therefore no longer unique in the weighted Euclidean model.

Young (1982) suggested fitting a weight matrix of some predetermined rank less than q. This implies that the private spaces have fewer dimensions than the stimulus space and thus not all subjects need to share the same set of axes.

Orthogonal decomposition of the weight matrix

5.3.1   Carroll and Chang (1972) applied the spectral decomposition to the weight matrix, giving $\mathbf{W}_s = \mathbf{P}_s \mathbf{\Lambda}_s \mathbf{P}'_s$, where the columns of $\mathbf{P}_s$ are the eigenvectors of $\mathbf{W}_s$ and $\mathbf{\Lambda}_s$ is the diagonal matrix of eigenvalues. Then the coordinates of the private spaces can be defined as

$$\mathbf{X}^{(s)} = (\mathbf{X}\mathbf{P}_s)\mathbf{\Lambda}_s^{1/2} \tag{5.18}$$

since

$$(\mathbf{x}_i - \mathbf{x}_j)'\mathbf{W}_s(\mathbf{x}_i - \mathbf{x}_j) = ((\mathbf{x}_i - \mathbf{x}_j)'\mathbf{P}_s\mathbf{\Lambda}_s^{1/2})(\mathbf{\Lambda}_s^{1/2}\mathbf{P}_s'(\mathbf{x}_i - \mathbf{x}_j))$$

$$= (\mathbf{X}_i^{(s)} - \mathbf{X}_j^{(s)})(\mathbf{X}_i^{(s)} - \mathbf{X}_j^{(s)})' \tag{5.19}$$

In equation (5.18), the multiplication by $\mathbf{P}_s$ can be interpreted as a rotation of the stimulus space (the columns of $\mathbf{P}_s$ are orthonormal) and this is followed by multiplication with the diagonal 'weight' matrix of the square roots of the eigenvalues $\mathbf{\Lambda}_s^{1/2}$, which corresponds to a stretching and shrinking of the new rotated dimensions. It is now also easy to see that arbitrarily rotating the group stimulus axes does not change the solution if the inverse rotation is applied to the rotation matrix $\mathbf{P}_s$.

Young (Young, 1982, 1987) calls the columns of the matrix $\mathbf{P}_s\mathbf{\Lambda}_s^{1/2}$ the *principal directions*. The principal directions can be displayed in the group stimulus space by arrows pointing from the origin in the direction of the corresponding column vector.

Oblique decomposition of the weight matrix

5.3.2   Tucker (1972) and Harshman (1972) suggested using an oblique decomposition of the weight matrix. The idea is to present the private spaces in the same coordinate system as the group stimulus space. However, the fact that $\mathbf{W}$ has non-zero off-diagonal elements means that the distance between two points in private space is not only a weighted sum of the squared differences in the components along the coordinate axes $(x_{ik} - x_{jk})^2$, but also depends on the cross products of these differences $(x_{ik} - x_{jk})(x_{il} - x_{jl})$. One way of interpreting this, is to think of the axes in subject space as being oblique as well as being differently scaled. Then $\mathbf{W}$ can be viewed as a covariance matrix of the private space axes. Instead of using the spectral decomposition, the weight matrix is factored as

$$\mathbf{W}_s = \mathbf{D}_s^{1/2}\mathbf{R}_s\mathbf{D}_s^{1/2} \tag{5.20}$$

where $\mathbf{D}_s$ is just a diagonal matrix with elements equal to the diagonal elements of $\mathbf{W}_s$ and $\mathbf{R}_s$ is a symmetric matrix having ones in the diagonal. This can be seen by considering element W_{ij} (dropping subscript s for clarity)

$$W_{ij} = \sum_k \sum_l D_{ik}^{1/2} R_{kl} D_{lj}^{1/2}$$

$$= D_{ii}^{1/2} R_{ij} D_{jj}^{1/2} \tag{5.21}$$

where the summations disappear because the off diagonal elements of $\mathbf{D}_s$ are all zero.

It is clear from the form of (5.20) that $\mathbf{R}_s$ is just the correlation matrix of the private space axes. In order to represent the personal space coordinates in stimulus space and get the closest match possible between the Euclidean distances in this space and those in the true private space with oblique axes, the dimensions need to be scaled by $\mathbf{X}_s = \mathbf{D}^{1/2}\mathbf{X}$. The true Euclidean distance between two points (taking account of the angle between the true axes) is then given by

$$(\mathbf{x}_i^{(s)} - \mathbf{x}_j^{(s)})'\mathbf{R}(\mathbf{x}_i^{(s)} - \mathbf{x}_j^{(s)}) \tag{5.22}$$

Example 5.4

The rotated and weighted Euclidean model (non-metric) has been fitted to the drinks data using the squared distance algorithm. The three subjects' private spaces are given in Fig. 5.9.

These configurations are not very easy to compare because of their different orientations. A better way of looking at individual differences is to display the principal directions from an orthogonal decomposition of the weight matrix as illustrated in Fig. 5.10. The principal direction arrows determine the transformation of the group stimulus space to the private space as follows. First the plot is rotated so that dimensions 1 and 2 are aligned with arrows 1 and 2 respectively and then the new dimensions are stretched proportionally to the arrows' lengths.

In Fig. 5.10, the first principal direction corresponds closely to dimension 2, the type of drink, and the second, less important, principal direction corresponds to dimension 1, the colour. This result is equivalent to that obtained using the weighted Euclidean model. Since the principal directions of the other two subjects also almost coincide with the stimulus space dimensions, we have not gained much by allowing the weight matrix to be non-diagonal.

Using the oblique decomposition, the correlations between subject space axes are 0.08, −0.05 and −0.04 respectively. This corresponds to angles between private space axes of 85.4°, 92.9° and 92.4° respectively (the correlation is equal to the cosine of the angle). As expected, these angles do not differ very much from 90°.   □

Procrustean individual differences

5.4 Procrustean individual differences scaling was developed by Lingoes and Borg (1978). The method starts with a separate scaling solution $\mathbf{X}_s$, for each subject. Using generalized Procrustes analysis as described in Chapter 4, these configurations are brought into maximum alignment with each other by rotation, translation and scaling and a centroid configuration $\mathbf{Z}$ is formed which represents the group stimulus space.

In order to model the individual differences, the group stimulus space $\mathbf{Z}$ is then transformed to match it as closly as possible to each subject's original configuration $\mathbf{X}_s$. One possible tranformation is to weight differentially the dimensions of $\mathbf{Z}$ and this is essentially equivalent to the weighted Euclidean model. The fit of the transformed stimulus space to the original configurations may then be compared with the fit of a series of more complex transformations in order to choose the most appropriate model for the data. The different transformations or models which are typically used fall into two groups, *dimension weighted models* (or distance models) and *vector weighted models* (or perspective models).

The dimension weighted models, usually referred to as P_1 and P_2 are

Subject 1

(i,j)	$d_{ij}^{(s)}$	$\hat{d}_{ij}^{(s)}$	$\delta_{ij}^{(s)}$
1	1.26	1.30	1
2	1.30	1.34	2
4	2.02	2.10	5
3	2.02	2.10	4
5	2.15	2.10	3
6	2.41	2.35	6

S-Stress = 0.062 Stress = 0.031 $r_1^2 = 0.985$

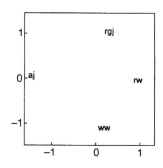

Subject 2

(i,j)	$d_{ij}^{(s)}$	$\hat{d}_{ij}^{(s)}$	$\delta_{ij}^{(s)}$
1	1.04	1.02	1
2	1.52	1.54	3
4	1.55	1.54	2
6	2.24	2.25	5
5	2.25	2.25	6
3	2.27	2.25	4

S-Stress = 0.013 Stress = 0.008 $r_2^2 = 0.999$

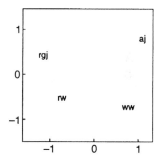

Subject 3

(i,j)	$d_{ij}^{(s)}$	$\hat{d}_{ij}^{(s)}$	$\delta_{ij}^{(s)}$
2	1.07	1.05	1
1	1.48	1.52	3
3	1.54	1.52	2
6	2.24	2.26	5
5	2.25	2.26	6
4	2.29	2.26	4

S-Stress = 0.023 Stress = 0.013 $r_3^2 = 0.997$

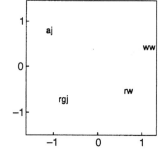

Figure 5.9 Rotated and weighted Euclidean model solution. The information given is the same as in Fig. 5.4.

equivalent to the weighted Euclidean model and the rotated and weighted Euclidean model respectively. In P_1, the dimensional salience model, the dimensions of the centroid configuration are differentially weighted giving a transformed centroid $\mathbf{Z}_s^* = \mathbf{Z}\mathbf{U}_s$, where $\mathbf{U}_s$ is a diagonal matrix of dimension weights. Note that the weights in this model correspond to the square roots of the weights $\mathbf{W}_s$ of the weighted Euclidean model except that they are now allowed to be negative giving rise to reflections as well as dilations. In P_1, one

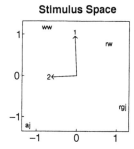
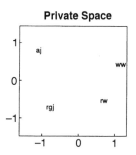

Figure 5.10   The group stimulus space with the principal directions for subject 3 and the private space for subject 3.

optimal rotation of $\mathbf{Z}$ is found for all subjects before multiplication by $\mathbf{U}_s$, whereas P_2, the individually rotated dimensional salience model, allows different rotations $\mathbf{Z}_s^r$ for each subject before the dimensions are weighted.

The vector weighted models allow the stimulus points of the centroid configuration to move radially away from or towards the origin. This is achieved simply by weighting the position vectors of the stimulus points using the transformation $\mathbf{Z}_s^* = \mathbf{V}_s\mathbf{Z}$ where $\mathbf{V}_s$ is an $(n \times n)$ diagonal matrix of weights. A negative weight has the effect of 'flipping' the vector in the opposite direction. The position of the origin, also called the *perspective point*, is as important in determining the configurations obtainable by the vector weighted models as is the orientation of the axes in determining configurations obtainable by the dimension weighted models. For the simple perspective model, P_3, one optimal perspective point (i.e. one translation of $\mathbf{Z}$) is found for all subjects before the transfomation is applied. The individually translated perspective model, P_4, allows subjects to have different translations of the centroid configuration, $\mathbf{Z}_s^t$, before the vector weighting is applied. The most complex model P_5 uses both vector weights and dimension weights and is given by $\mathbf{Z}_s^* = \mathbf{V}_s\mathbf{Z}\mathbf{W}_s$. The simplest model is P_0 with $\mathbf{Z}_s^* = \mathbf{Z}$, the result of generalized Procrustes analysis.

The six models are estimated by minimizing the sum of squared differences between corresponding coordinates of the transformed centroid $\mathbf{Z}_s^*$ and similarity transformations $\tilde{\mathbf{X}}_s$ of the individual configurations

$$L = \sum_s \text{tr}\{(\tilde{\mathbf{X}}_s - \mathbf{Z}_s^*)(\tilde{\mathbf{X}}_s - \mathbf{Z}_s^*)'\} \qquad (5.23)$$

Initially, when deriving $\mathbf{Z}$ for model P_0, the origins of $\tilde{\mathbf{X}}_s$ are placed at the means and the scales are normed by setting $\text{tr}(\tilde{\mathbf{X}}_s\tilde{\mathbf{X}}_s') = 1$. (This scale of $\tilde{\mathbf{X}}_s$ remains fixed for all models.)

For the dimension weighted models (given by $\mathbf{Z}_s^* = \mathbf{Z}\mathbf{U}_s$ or $\mathbf{Z}_s^* = \mathbf{Z}_s^r\mathbf{U}_s$) L is minimized with respect to the dimension weights $\mathbf{U}_s$ and the rotation(s) of $\mathbf{Z}_s^*$. The translation of $\tilde{\mathbf{X}}_s$ and $\mathbf{Z}_s^*$ remains as in model P_0 and the rotation of $\tilde{\mathbf{X}}_s$ is the same as that of $\mathbf{Z}_s^*$. For the vector weighted models (given by $\mathbf{Z}_s^* = \mathbf{V}_s\mathbf{Z}$ or $\mathbf{Z}_s^* = \mathbf{V}_s\mathbf{Z}_s'$) L is minimized with respect to the vector weights $\mathbf{V}_s$ and the translation(s) of $\mathbf{Z}_s^*$, as well as with respect to the translation and rotation of $\tilde{\mathbf{X}}_s$. The rotation of $\mathbf{Z}_s^*$ is arbitrary.

Table 5.4 The Procrustean individual differences models and the number of parameters estimated per subject

Model	Transformation Z_s^*	Number of parameters per subject
P_0	Z	no parameters
P_1	ZU_s	q dimension weights
P_2	$Z_s'U_s$	q dimension weights and $\binom{p}{2}$ rotation parameters
P_3	V_sZ	n vector weights
P_4	$V_sZ_s^t$	n vector weights and q translation parameters
P_5	V_sZU_s	q dimension weights and n vector weights

The strategy in Procrustean individual differences scaling is to fit all six models and to choose the best one by considering the goodness of fit of each of the models. The goodness-of-fit measure is defined as the squared correlation between the coordinates Z_s^* and $\tilde{X}_s$

$$g_s = r^2(Z_s^*, \tilde{X}_s) \tag{5.24}$$

which represents the proportion of variation in the individual configuration explained by the model. A combined goodness-of-fit measure is

$$S = \frac{1}{m}\sum_s (1 - g_s)^{1/2} \tag{5.25}$$

One difficulty with selecting one of the six models is the drastic increase in the number of parameters estimated per subject from model P_0 to model P_5 as shown in Table 5.4 and the associated inevitable increase in the goodness of fit. There is no statistical test to help decide on the best trade-off between parsimony and goodness of fit (but see Langeheine, 1982 for a statistical evaluation of measures of fit).

Sometimes one may be interested not in the way subjects' perceptions differ from each other but in the way they differ from a hypothesized structure. For example one may be interested in the way subjects' perception of the relative locations of cities differs from the true map. This analysis can be done by specifying the stimulus configuration Z externally instead of estimating it from the centroid of the individual configurations.

Example 5.5

We will use this method to analyse data from ten subjects who judged the distances between ten American cities. The data are row conditional, each row representing the ranked distances of all cities from a reference city, similarly to the body parts data in Table 5.3. The separate scaling solutions X_s were derived using a non-metric row conditional analysis. A geographical map derived from the flight distances between cities was used as the externally specified configuration Z. It can be seen in Table 5.5 that the fit of the simplest model, corresponding here to Procrustes alignment of the individual maps to the hypothesized map, is already very good, the average percentage of variation in the coordinates explained by the model being 91%.

Table 5.5   Percentage of variance explained by the six models fitted to the city data

Subject	Model Z_s^*					
	Z	ZU_s	$Z_s'U_s$	V_sZ	V_sZ_s'	V_sZU_s
1	94	94	95	98	100	97
2	89	90	91	93	99	92
3	93	94	94	95	99	95
4	95	96	96	98	99	98
5	90	91	92	95	97	94
6	88	89	90	95	99	95
7	86	87	87	95	99	90
8	96	96	96	99	100	98
9	91	91	91	98	98	92
10	82	87	87	91	96	91
Mean	91	92	92	96	99	94

The dimension weighted model accounts for another 1% of the variation with an increase by two parameters per person and the vector weighted model for another 5% with an increase by ten parameters per person. Without deciding which is the best model, we use this opportunity to illustrate the vector weighted model. In Figs 5.11(a) and (b), the hypothesis configuration and configuration 7 are shown in the orientation which was optimal for the dimension weighted model P_2. Figure 5.11(c) shows how the hypothesis configuration is vector weighted to match configuration 7 using the perspective point (common to all subjects) indicated by a '+'. In Fig. 5.11(d), the vector weighted hypothesis configuration has been rotated and translated (circles) to optimally fit configuration 7 (diamonds). (Note that the algorithm rotates and translates the subject's configuration not the centroid.) The city requiring the greatest shift is Denver, followed by San Francisco, two cities that were very badly placed in configuration 7.

The distance by which a city moves away from the perspective point is given by the simple formula (vector weight-1) × (distance of point from perspective point) with a negative distance indicating a move towards the perspective point. These distances are given in Table 5.6 for all subjects together with the city or state the subjects came from.

It will be left as an exercise for the reader to determine if there is any relationship between a subject's place of origin and their misjudgement of the various cities' relative locations.   □

Inference in multidimensional scaling

5.5   The scaling methods described for a single proximity matrix in Chapter 3 and the three-way methods described in this chapter have all been essentially 'descriptive'; none incorporates an estimation of variability of points in an observed configuration, so that 'inferences' drawn from such configurations have, up to now, been relatively informal. In a series of papers, Ramsay (1977, 1978a, 1982) has attempted to confront this problem by postulating

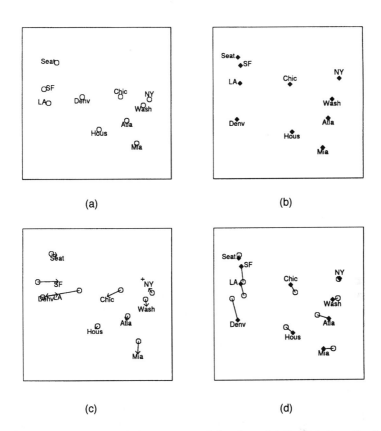

(a)　　　　　　　　　　(b)

(c)　　　　　　　　　　(d)

Figure 5.11 Illustration of the vector-weighted model P_3; (a) hypothesis configuration; (b) configuration of subject 7; (c) arrows from hypothesis configuration to vector-weighted hypothesis configuration for subject 7 with the perspective point marked as +; (d) configuration for subject 7 (diamonds) and Procrustes rotated vector-weighted hypothesis configuration for subject 7 (circles).

Table 5.6 The distances by which the vector weights move cities away from the perspective point (north east, near New York) for subjects from different places of origin. (Negative values indicate moving towards the perspective point)

Place of origin	Hous	Seat	Wash	Mia	Chic	LA	SF	Denv	NY	Atla
Illinois	−13	2	−5	−7	5	4	−5	13	−3	−6
Chicago	−19	2	−10	1	4	−2	−5	3	−9	−4
Nebraska	−5	−6	−1	9	0	2	−7	−4	−1	0
Los Angeles	−7	−4	−2	10	−5	3	−9	3	−3	−4
Los Angeles	−13	−3	−4	5	−2	0	−11	11	−4	−8
New Jersey	−15	2	−10	−8	18	6	−5	8	−13	−10
New York	2	−4	5	8	10	−10	−14	23	−4	3
Connecticut	−1	2	−12	0	10	2	−3	7	−7	−8
Florida	−12	−2	−13	2	1	−1	−3	9	−11	−26
North Carolina	−2	−4	−4	10	15	−3	−13	5	−12	−12

distributional models for observed dissimilarities that allow the use of likelihood methods in MDS, with the subsequent possibility of confidence regions for points in the resulting solutions. Other authors have developed alternative inferential methods (see for example Takane and Sergent, 1983; Zinnes and MacKay, 1983) but here we will describe only the approach suggested by Ramsay.

Ramsay (1982) assumes that the true dissimilarities $d_{ij}^{(s)}$ correspond exactly to the distances of the weighted Euclidean model

$$d_{ij}^{(s)} = \left\{ \sum_k w_k^{(s)} (x_{ik} - x_{jk})^2 \right\}^{1/2}$$

for some true configuration of points and set of weights. The observed transformed dissimilarities (disparities)

$$\hat{d}_{ij}^{(s)} = f^{(s)}(\ln(\delta_{ij}^{(s)})) \tag{5.26}$$

are independently distributed as

$$\hat{d}_{ij}^{(s)} \sim N(\ln(d_{ij}^{(s)}), (\sigma_{ij}^{(s)})^2) \tag{5.27}$$

The reason for the log transformation is that the dissimilarity judgements tend to become less accurate as the true dissimilarity increases.

The variance in (5.27) is modelled as

$$(\sigma_{ij}^{(s)})^2 = \sigma_s^2 \gamma_{ij}^2 \tag{5.28}$$

where the term σ_s^2 is the subject specific variance component and allows for some subjects to be more precise than others, possibly because they may be more motivated or more familiar with the set of stimuli. The term γ_{ij}^2 represents the stimulus pair-specific variance component. This term can be further decomposed into

$$\gamma_{ij}^2 = \frac{1}{2}(a_i^2 + a_j^2), \quad \sum a_i^2 = n \tag{5.29}$$

where a_i and a_j represent the relative precision of judgements involving stimuli i and j respectively.

Ramsay suggests three different transformations from dissimilarities to disparities. The simplest one is the scale transformation

$$f^{(s)}(\ln \delta_{ij}^{(s)}) = \ln \delta_{ij}^{(s)} + \nu_s$$

$$= \ln(\delta_{ij}^{(s)} e^{\nu_s}) \tag{5.30}$$

which allows the disparities of different subjects to be scaled by different multiplicative constants before taking the logarithm. The power transformation

$$f^{(s)}(\ln \delta_{ij}^{(s)}) = p_s \ln \delta_{ij}^{(s)} + \nu_s$$

$$= \ln[(\delta_{ij}^{(s)})^{p_s} e^{\nu_s}] \tag{5.31}$$

additionally allows different powers of the dissimilarities to be taken before taking logarithms. This captures a certain type of non-linearity between dissimilarities and distances. Ramsay also suggests a more general nonlinear function between distances and dissimilarities given by

$$f^{(s)}(\ln \delta_{ij}^{(s)}) = S^{(s)}(\ln \delta_{ij}^{(s)}) + \nu_r \tag{5.32}$$

where S is a monotonic spline transformation. This transformation may be useful when the data are ordinal.

The configuration $\mathbf{X}$, the weights $w_k^{(s)}$, the transformation parameters and variances are estimated by maximizing the log likelihood of the log dissimilarities. The log likelihood value can then also be used to select the best model using asymptotic theory.

We will now derive the log likelihood for the model using the power transformation in (5.31). The model can be written as

$$p_s \ln \delta_{ij}^{(s)} + \nu_s = \ln d_{ij}^{(s)} + \epsilon_{ij}^{(s)} \tag{5.33}$$

so that, for independently normally distributed errors $\epsilon_{ij}^{(s)}$, the probability density function of the log dissimilarities is given by

$$\ln \delta_{ij}^{(s)} \sim \frac{p_s}{\sigma_{ij}^{(s)} \sqrt{2\pi}} \exp\left(-\frac{1}{2} \left(\frac{\epsilon_{ij}^{(s)}}{\sigma_{ij}^{(s)}} \right)^2 \right) \tag{5.34}$$

and the log likelihood of the log dissimilarities is given by

$$L = c + \sum_s \left\{ M_s \ln p_s - \sum_{ij} \ln \sigma_{ij}^{(s)} - \frac{1}{2} \sum_{ij} \left(\frac{\epsilon_{ij}^{(s)}}{\sigma_{ij}^{(s)}} \right)^2 \right\} \tag{5.35}$$

where c is a constant and M_s is the number of stimuli for which subject s judged the similarities. The solution is found in an iterative algorithm involving four stages. First, L, is maximized with respect to the stimulus configuration $\mathbf{X}$ keeping all other parameters fixed at their current estimates. Secondly, L is maximized with respect to the weights $w_k^{(s)}$, again keeping all other parameters fixed, thirdly with respect to the transformation parameters p_s and ν_s and finally with respect to the variances $(\sigma_{ij}^{(s)})^2$. (Details of the algorithm are given in Ramsay, 1977.) Note that, as Ramsay (1991) points out, the variances $(\sigma_{ij}^{(s)})^2$ are estimated assuming all other parameters to be fixed rather than estimated, which leads to a downward bias. For small numbers of degrees of freedom, Ramsay (1980) suggests a correction factor for the χ^2 tests of dimensionality derived from simulation studies.

Example 5.6

As an example of Ramsay's approach, we will apply his algorithm to his data on the dissimilarities of 14 emotions by 10 subjects. The data for the first two subjects are presented in Table 5.7.

For this analysis, the power transformation and lognormal distribution were used and separate variance components were estimated for different subjects assuming that $(\sigma_{ij}^{(s)})^2 = \sigma_s^2$. (Estimating variance components for all pairs of stimuli would require a much larger sample size.) Two models were estimated, the weighted Euclidean model and the weighted Euclidean model with weights

Table 5.7 Dissimilarity ratings (from 1 to 9) of 14 emotions by subject 1 (lower diagonal) and subject 2 (upper diagonal). The emotions are: satisfied, fascinated, surprised, eager, happy, passionate, affectionate, despising, panicky, afraid, guilty, sad, angry and rejected

	sat	fas	sur	eag	hap	pas	aff	des	pan	afr	gui	sad	ang	rej
sat		3	4	6	1	2	3	9	8	9	8	8	8	9
fas	7		5	2	3	4	2	9	9	8	7	7	8	9
sur	7	7		5	3	5	5	7	3	7	7	5	6	7
eag	9	4	9		6	1	2	7	9	6	9	6	7	7
hap	1	3	6	9		2	4	9	8	9	7	9	9	8
pas	3	3	4	2	1		2	1	5	8	6	5	2	8
aff	6	2	9	6	7	1		9	7	8	7	7	9	9
des	9	7	8	9	9	3	9		5	4	4	3	2	2
pan	9	6	4	2	9	6	6	9		3	6	4	3	5
afr	9	6	3	9	9	6	9	4	1		3	4	3	5
gui	9	9	9	9	9	3	9	9	2	4		4	4	4
sad	9	7	3	8	9	3	6	9	2	2	4		2	2
ang	9	9	7	6	9	1	5	2	4	3	9	6		2
rej	9	9	7	9	9	9	9	9	2	1	4	1	3	

constrained to equal one. (The latter is the replicated multidimensional scaling method mentioned in Section 5.1). There were 910 data points ($7 \times 13 \times 10$) and the models required 63 and 53 parameters respectively, leaving 857 and 847 degrees of freedom. The log likelihoods differed only by 10 so that the replicated model was selected.

The two-dimensional stimulus configuration for the replicated model is shown together with approximate 99% confidence regions in Fig. 5.12.

It is very clear that similar emotions such as 'satisfied' and 'happy' are close together and that dimension 1 separates positive from negative emotions. The various parameters fitted for the ten subjects are given in Table 5.8.   □

Ramsay's more formal approach to multidimensional scaling has not been universally welcomed. In the discussion following Ramsay (1982), several individuals were sceptical about the need for inferential multidimensional scaling in general and about the assumptions underlying Ramsay's model in particular. Chatfield, for example doubted that 'there is indeed a true point for each stimulus' and worried that the 'model is too complicated and still not realistic enough'. In particular he was concerned with the assumption that the residuals are independent. Similarly, Silverman professed to being a 'little uneasy about the use of multidimensional scaling as a model-based inferential technique, rather than just an exploratory or presentational method'. He further stated that he 'finds it hard to accept the assumptions behind the model'. De Leeuw feared that programs such as that written by Ramsay to implement his method 'tend to induce a false sense of confidence in their users, by providing interval estimates and tests of hypotheses which may actually be quite useless or misleading' when the assumptions of the model are violated. For similar reasons, Kruskal and Carroll (1969) defend the exploratory approach to multidimensional scaling by stating; 'We believe that the use of badness-of-fit

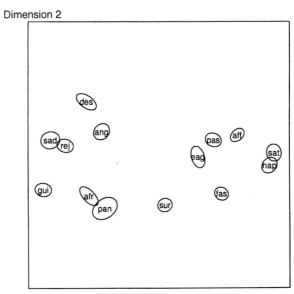

Figure 5.12 Two-dimensional configuration for the ratings of dissimilarity of emotions showing approximate 99% confidence intervals.

Table 5.8 Dimension weights, power transformation and variances for the ten subjects

	ν_s	p_s	σ_s
Subject 1	0.15	1.46	0.76
Subject 2	0.30	1.42	0.50
Subject 3	0.58	1.32	0.38
Subject 4	−0.30	1.70	0.62
Subject 5	−0.14	1.68	0.63
Subject 6	0.81	1.31	0.66
Subject 7	0.95	1.20	0.65
Subject 8	−1.21	2.28	0.64
Subject 9	−0.13	1.67	0.70
Subject 10	−1.01	2.07	0.66

functions, where no stochastic element is explicitly present in the model, is a perfectly well-justified procedure when greater detail and sophistication is either impractical or unjustified by the current state of knowlege'.

Several discussants of Ramsay (1982) are also worried about the validity of asymptotic theory and the stability of the solution. Chatfield comments that 'procedures based on asymptotic theory can go wildly wrong if many parameters are used' while Coxon warns that enclosing point locations with confidence regions may mislead the user 'into thinking that the problem of local instability of point location in MDS configurations has been effectively solved'.

Storms (1995) has investigated the robustness of Ramsay's approach to departures of the data from the assumed lognormal error distribution. Data were simulated using normal, rectangular and lognormal error distributions with various error variances and the author concludes that the error distribution has virtually no effect on the estimated distance parameters. Further, Ramsay's tests of dimensionality (corrected for small samples) were quite robust to violations of the error model.

Many of the criticisms of Ramsay's approach may have some weight but from a pragmatic point of view we feel his method may still provide a valuable tool for strengthening the usual informal scaling approach. Perhaps critics should remember that well known aphorism 'all models are wrong but some are useful'.

Summary

5.6 Individual differences in dissimilarity data from different sources (for example different subjects) can be explored by deriving a group stimulus configuration and estimating the parameters of the geometric transformations required to match this group configuration as closely as possible to the individual dissimilarity matrices. The parameters of the transformation can be used to summarize the individual differences. Useful geometric transformations include dimension weighting, which may be preceded by individual rotations, and vector weighting, which may use different perspective points. Several different approaches to the problem of estimating the group stimulus configuration and transformation parameters have been described including an inferential method.

6

Asymmetric and Rectangular Data

Introduction

6.1 The scaling methods discussed thus far are only suitable for the analysis of symmetric proximity matrices. This covers the large majority of practical applications, but there are, however, situations in which the observed similarities or dissimilarities are asymmetric. An example of such a situation involving judgements of the similarity of Morse Code signals was, in fact, introduced in Chapter 1. In Table 1.3, the percentage of times the pair of signals i and j were judged to be the same differed when the signals were presented in the order i followed by j rather than j followed by i. Other examples of where asymmetry occurs in practice are the number of marriages between men of one nationality and women of another, the time taken to travel between two locations, particularly in a hilly area, diallel-cross experiments and immigration/emigration statistics. Asymmetric matrices of this type could be dealt with by the usual scaling procedures by simply using only the lower or upper triangular matrix, or alternatively, their average. A further possibility that essentially ignores asymmetry is to derive a solution in which the distance between stimuli i and j is as close as possible to both disparities $\hat{d}_{ij}$ and $\hat{d}_{ji}$, by defining stress to involve summing over all elements of the disparity matrix, not simply the upper or lower triangular part. All such approaches are essentially assuming that the differences between the upper and lower triangular parts of the matrix are simply due to 'error' or 'noise'.

However, in many situations, the asymmetry itself is at least as important as the symmetric features of the matrix. For example, in immigration/emigration data, it would clearly be of interest if migration from A to B was substantially greater than from B to A. Several methods have been developed for modelling and displaying the asymmetry in square asymmetric matrices.

A more general approach to analysing an asymmetric matrix would be to view its rows and columns as corresponding to different stimuli and to seek a configuration containing $2n$, q-dimensional points $\mathbf{X}(n \times q)$ for the rows and $\mathbf{Y}(n \times q)$ for the columns. The amount of asymmetry between stimuli i and j

could then be judged by the difference between the distances x_i to y_j and y_i to x_j. The general procedure for constructing such a configuration is *multidimensional unfolding*. In fact, this is a general method which can be applied also to rectangular matrices in which the rows and columns really do refer to different objects, the two-way, two-mode data described in Chapter 1. Multidimensional unfolding and other methods for such rectangular matrices will be discussed in Section 6.3. Here, however, we continue with an account of methods specifically for the analysis of square asymmetric matrices.

The analysis of square asymmetric matrices

6.2 Gower (1977) and Constantine and Gower (1978, 1982) proposed a method for scaling asymmetric matrices that relies essentially on the canonical decomposition of skew symmetric matrices. Any asymmetric matrix can be written as the sum of a symmetric matrix **M** and a skew symmetric matrix **N**

$$\mathbf{D} = (\mathbf{D} + \mathbf{D}')/2 + (\mathbf{D} - \mathbf{D}')/2$$

$$= \mathbf{M} + \mathbf{N} \tag{6.1}$$

where **M** and **N** are orthogonal to each other so that the total sum of squares of **D** is simply the sum of the sums of squares of **M** and **N**. The symmetric features of the data can be dealt with by scaling **M** using any of the methods described in Chapter 3. The elements n_{ij} of **N** represent the amount by which d_{ij} differs from the mean $m_{ij} = (d_{ij} + d_{ji})/2$, so that $n_{ij} = -n_{ji}$. The asymmetry of **D** can be analysed by decomposing **N** using the canonical decomposition of skew-symmetric matrices, i.e.

$$\mathbf{N} = \sum_{k=1}^{[n/2]} \sigma_{2k} \left(\mathbf{u}_{2k-1} \mathbf{u}'_{2k} - \mathbf{u}_{2k} \mathbf{u}'_{2k-1} \right) \tag{6.2}$$

where $[n/2]$ is the largest integer, less than or equal to $n/2$ and σ_k and $\mathbf{u}_k$ are the eigenvalues and eigenvectors of $\mathbf{NN}'$ respectively. The best rank 2 approximation to **N** is given by the first term in (6.2) which is proportional to $\mathbf{u}_1 \mathbf{u}'_2 - \mathbf{u}_2 \mathbf{u}'_1$. A two-dimensional map can be produced by plotting the points (u_{1i}, u_{2i}), $i = 1, \ldots, n$. Then n_{ij} is approximately proportional to $(u_{1i}u_{2j} - u_{2i}u_{1j})$, the cross-product of the position vectors of points i and j and hence the area of the parallelogram (equivalently the triangle) spanned by these vectors.

Example 6.1

We will illustrate this method using the data shown in Table 6.1, adapted from Hartigan (1975). The data give the percentage of people in various European countries who claim to speak the other European languages enough to make themselves understood. The element n_{ij} of the skew symmetric matrix for these data represents (half) the difference between the percentage of people from a country i who speak language j and the percentage of people from country j who

Table 6.1 Number of persons (%) speaking a language 'enough to make yourself understood' (from Hartigan, 1975).

Country	German	Italian	French	Dutch	English	Portuguese	Swedish	Danish	Norwegian	Finnish	Spanish
West Germany	100	2	10	2	21	0	0	0	0	0	1
Italy	3	100	11	0	5	0	0	0	0	0	1
France	7	12	100	1	10	1	2	3	0	0	7
Netherlands	47	2	16	100	41	0	0	0	0	0	7
Great Britain	7	3	15	0	100	0	0	0	0	0	2
Portugal	0	1	10	0	9	100	0	0	0	0	2
Sweden	25	1	6	0	43	0	100	10	11	5	1
Denmark	36	3	10	1	38	0	22	100	20	0	1
Norway	19	1	4	0	34	1	25	19	100	0	0
Finland	11	1	2	0	12	0	23	0	0	100	0
Spain	1	2	11	0	5	0	0	0	0	0	100

Dimension 2

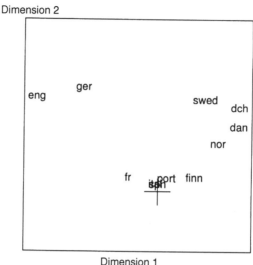

Dimension 1

Figure 6.1 Canonical decomposition of language data in Table 6.1. The cross indicates the origin.

speak language *i*. For example for Germany and Sweden, the percentage of Germans who speak Swedish is 0% and the percentage of Swedes who speak German is 25% so that the corresponding elements in the skew-symmetric matrix are ±12.5%. The canonical decomposition of the skew-symmetric matrix is given in Fig. 6.1.

This two-dimensional representation accounts for 94% of the sum of squares of the skew-symmetric matrix. The substantial asymmetry between Swedish and German is reflected by the area of the triangle formed by connecting the points for Sweden and Germany to each other and to the origin, indicated by a cross. From the figure it is clear that the asymmetry between English and Dutch is even greater. The languages Italian, Spanish, Portuguese, Dutch, Danish and Norwegian are close to a straight line from the origin so that none of the triangles between these countries has a large area. This implies that the submatrix formed by these countries is almost symmetrical. Spanish and Italian are very close to the origin, giving small triangles with all other languages, implying that the percentages of Italians and Spaniards speaking other languages is not very different from those of other countries speaking Spanish and Italian. □

Permutation method

6.2.1 An alternative method of analysing asymmetric proximity matrices proposed by Gower (1977) is to permute the rows and columns of **D** until the absolute difference between the sums of the elements of the upper and lower triangular matrices is a maximum. This permutation gives the worst possible asymmetry which may then be analysed by comparing the maps derived for the upper and lower triangular matrices.

Table 6.2 Permutation of language data maximizing the difference between the upper and lower triangular matrices.

Country	French	English	German	Italian	Spanish	Dutch
France	0	90	93	88	93	99
Great Britain	85	0	93	97	98	100
Germany	90	79	0	98	99	98
Italy	89	95	97	0	99	100
Spain	89	95	99	98	0	100
Netherlands	84	59	53	98	98	0

Example 6.2

A submatrix was formed from Table 6.1 for the languages German, Italian, French, Dutch, English and Spanish. Dissimilarities were formed by subtracting the percentages from 100%, giving the percentage of people in each country who do not speak a given language. The permutation of languages giving the worst asymmetry is given in Table 6.2.

Note that the permutation itself already reveals useful information about the asymmetry. For example, the fact that French comes first and Dutch last means that the upper triangular matrix contains the percentages of French people who do not speak each of the other languages (first row) and the percentages of people in each of the other countries who do not speak Dutch (last column). These percentages are all large compared with the percentages (in the lower diagonal matrix) of people in all other countries who do not speak French (first column) and of people in the Netherlands who do not speak each of the other languages (last row). The classical multidimensional scaling solutions for the upper and lower triangular matrices have been rotated to best fit (see Procrustes rotation, Chapter 4) and are presented in Fig. 6.2.

Although the fit of the scaling solutions is not very good (stress = 0.36 and 0.41), certain features are still apparent. For example, in the map of the lower triangular matrix (left), Dutch, German and English are close to each other because many Germans and Dutch speak English and many Dutch speak

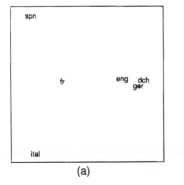
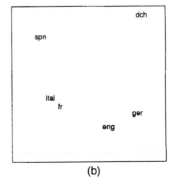

(a)   (b)

Figure 6.2 Classical multidimensional scaling solution of (a) lower triangular matrix and (b) upper triangular matrix in Table 6.2.

German. In the map of the upper triangular matrix (right) however, these three countries are far apart because not many British speak German or Dutch and not many Germans speak Dutch.　□

First way weights

6.2.2　A further method for dealing with asymmetric proximity matrices, which is more in line with the scaling procedures encountered in earlier chapters, was proposed by Young (1975). The model is very similar to the weighted Euclidean model described in Chapter 5. The solution consists of an overall stimulus configuration $\mathbf{X}$ which summarizes the symmetric features of the matrix, as well as a set of n diagonal matrices of weights, $\mathbf{V}_i$ ($q \times q$), one for each row stimulus. The dissimilarities between the ith row stimulus and all the column stimuli are approximated by the distances

$$d_{ij} = [\sum(\mathbf{x}_i - \mathbf{x}_j)'\mathbf{V}_i(\mathbf{x}_i - \mathbf{x}_j)]^{1/2}, \quad j = 1,\ldots,n \tag{6.3}$$

and separate configurations can be derived for each row object by scaling the dimensions as

$$\sqrt{\mathbf{V}_i}\mathbf{X}'$$

The configuration derived using the weight matrix $\mathbf{V}_i$ for row object i reflects the relationship between that row object and all the column objects, i.e. the distances $\{d_{ij}, j = 1,\ldots,n\}$ approximate the dissimilarities $\{\delta_{ij}, j = 1,\ldots,n\}$. Note however that the distances between any other row objects and any of the stimuli, d_{kj}, $k \neq i$, have no interpretation in the configuration given by $\sqrt{\mathbf{V}_i}\mathbf{X}'$.

Example 6.3

We apply the model to the submatrix of the language data given in Table 6.2. The fit in two dimensions is good (S-Stress $= 0.07$, $r^2 = 0.98$). The simulus configuration is given in Fig. 6.3 together with the weighted solutions for French, English and Dutch.

Each weighted solution represents only the relationship between one country and all the languages. The relevant reference country in each configuration has been enclosed by a square. It is clear from the figure that the Dutch speak most languages better than the French and British.　□

A vector model

6.2.3　A very different approach to exploring asymmetry was proposed by Harshman *et al.* (1982) who describe the following decomposition of an asymmetric proximity matrix

$$\mathbf{D} = \mathbf{ARA}' + \mathbf{E} \tag{6.4}$$

This decomposition is analogous to a factor analysis with $\mathbf{A}$ representing the matrix of factor loadings and $\mathbf{R}$ the correlation matrix of the factors. One

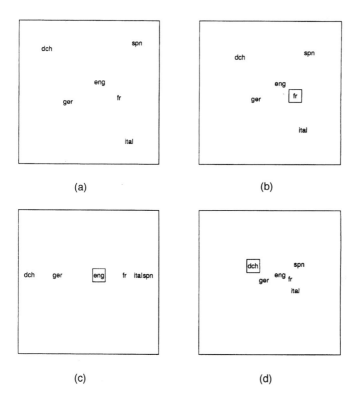

Figure 6.3 First way weights model applied to the submatrix in Table 6.2; (a) stimulus configurations; (b), (c) and (d) weighted solution for France, Britain and the Netherlands respectively.

important difference is that **R** is asymmetric; this matrix may be interpreted as describing the asymmetric relationship between the factors corresponding to the columns of **A**.

Example 6.4

Harshman *et al.* (1982) apply this decomposition to a matrix of free word association frequencies obtained as follows. The subjects were asked which phrase they thought of when the interviewer called out each of eight different phrases describing the quality of hair. The elements of the (8 × 8) matrix, given in Table 6.3 are the number of times each stimulus phrase by the interviewer evoked each of the other phrases in the respondent.

   The two-dimensional decomposition of this word association matrix is shown in Table 6.4. The two columns of **A** can be interpreted as representing the factors 'thickness' and 'vigour' and the (2 × 2) matrix **R** is roughly proportional to the frequency with which a phrase corresponding to 'thickness' evokes a phrase associated with 'vigour' etc.   □

Table 6.3 Word association frequencies: the number of times a stimulus phrase evoked each phrase.

Stimulus phrase	body	fullness	holds set	bouncy	not limp	manageable	zesty	natural
body		44	5	23	1	19	1	3
fullness	22		5	3	1	9	1	2
holds set	17	21		5	0	17	0	5
bouncy	15	12	3		1	5	0	14
not limp	28	27	4	18		4	1	7
manageable	17	13	11	2	0		0	3
zesty	7	9	2	22	0	4		13
natural	4	9	1	2	0	7	1	

Table 6.4 The **A** and **R** matrices for the word association data.

	Thickness	Vigour
Matrix A		
body	0.299	0.252
fullness	0.355	−0.158
holds set	0.041	0.213
bouncy	0.172	0.004
not limp	−0.048	0.420
manageable	0.150	0.013
zesty	−0.043	0.248
natural	0.074	0.010
Matrix R		
thickness	248	44
vigour	216	24

Explicit models

6.2.4   The methods for asymmetric matrices described in the previous sections are useful tools for visualizing asymmetry. However, in some cases we may want to model the asymmetry explicitly. A review of such models, is given by Zielman and Heiser (1996) who describe a general class of models called *similarity and bias* models. For similarity and bias models, the proximity of stimulus i to stimulus j is modelled as

$$\delta_{ij} = F(s_{ij} + r_i + c_j) + \epsilon_{ij} \tag{6.5}$$

where F is a general monotonic function, s_{ij} is a symmetric similarity function and r_i and c_j are bias functions on the rows and columns respectively. Here we will briefly describe one specific similarity and bias model, the distance model of Weeks and Bentler (1982). In distance models, the symmetric similarity function s_{ij} corresponds simply to the set of (usually Euclidean) distances d_{ij} between stimuli represented as points in a small number of dimensions. The model proposed by Weeks and Bentler is given by

$$\delta_{ij} = b d_{ij} + c_i - c_j + k + \epsilon_{ij} \tag{6.6}$$

The term $c_i - c_j$ $(= (\delta_{ij} - \delta_{ji})/2$, apart from errors) represents the skew symmetric part of the model and the constants c_i can be thought of as the 'dominance' or 'utility' values for stimuli. For example, when applied to journal citation data, c_j measures the surplus of citations received compared with citations given. A physical model which could give rise to the relationship in (6.6) is the jet stream model by Constantine and Gower (1978). Let the observed dissimilarities be equal to the time taken to fly from city i to another city j at constant air speed V when there is a wind of speed v flowing at an angle θ_{ij} to the line connecting the two cities. When v is small compared to V, Constantine and Gower (1978) show that the symmetric part of the model is approximately equal to the time taken to travel from city i to j in the absence of any wind and the skew symmetric part is given by

$$n_{ij} = (d_i - d_j) \frac{v}{V}$$

where d_i and d_j are the projections of the positions of cities i and j onto the direction of the wind. Weeks and Bentler (1982) estimate the parameters of the model in (6.6) by minimizing the sum for squared residuals using a steepest descent algorithm.

The analysis of rectangular matrices

6.3   Rectangular matrices of quantitative elements include the usual multivariate data matrix in which each of m individuals is observed by a set of n variables, contingency tables and preference data where each row (say) represents the scores (or perhaps ranks) from a particular judge for the set of stimuli. Preference data may also be collected by presenting subjects with each pair of stimuli and asking them which of the two stimuli in the pair they prefer. Such paired comparison data may be converted to rectangular preference data as described in Carroll (1972). (Methods specifically for pairwise preference data will not be discussed here; see for example De Soete and Carroll, 1983; De Soete et al., 1986.)

We will assume for now that the rows represent subjects and the columns are 'stimuli'. Following the notation in Chapter 5, we will use the subscripts s and i for subjects and stimuli respectively. The aim of multidimensional scaling is then to display both the subjects (rows) and the stimuli or variables (columns) as points $\mathbf{y}_s, s = 1, \ldots, m$ and $\mathbf{x}_i, i = 1, \ldots n$ in some space so that the relationship between subjects and stimuli implied by the observed preference matrix is represented adequately by the distances (usually Euclidean) between the subject and stimulus points in the derived configuration.

Distance models: multidimensional unfolding

6.3.1   The problem was first attacked by Coombs (1950) who introduced *unfolding* as a way of representing judges and stimuli on a single straight line

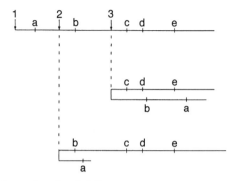

Figure 6.4   Unidimensional unfolding.

so that the rank order of the stimuli as determined by each judge is reflected by the rank ordering of the distances of the stimuli to that judge. The example in Fig. 6.4 illustrates how the name 'unfolding' arises. The line has been folded at the points for each judge so that the rank order of the stimuli can be read off from left to right. For example Judge 1 ranks the objects in the order a b c d e, Judge 2 in the order b a c d e and Judge 3 in the order c d b e a. In q dimensions, the folding corresponds to picking up a q-dimensional handkerchief at the point of each judge and folding it into a line.

Early work on fitting the unfolding model concentrated on finding exact solutions if they existed. Coombs (1950) solved the problem for non-metric unidimensional unfolding and Bennett and Hayes (1960) solved the non-metric multidimensional unfolding model. Finally, Schönemann (1970) developed an algorithm for finding the exact solution in metric multidimensional unfolding. Several authors consider probabilistic approaches in order to find approximate solutions when no exact solutions exist. Zinnes and Griggs (1974) for example, assume that the coordinates of the subject and stimulus points in the unfolding model are independently normally distributed and maximize the likelihood of the observed rankings (for non-metric unfolding) with respect to the coordinate means (for fixed coordinate variances). For more details of these methods see the original articles or Cox and Cox (1994).

These early approaches have been replaced by methods which aim to minimize a loss function in very much the same way as the scaling methods discussed in Chapters 3 and 5. The unfolding model is analogous to the multidimensional scaling model except that there are separate points for row and column objects and it is the distances between row and column objects (or their rank ordering) that approximate the original data. We have

$$\hat{d}_{si}^2 = (\mathbf{x}_i - \mathbf{y}_s)'(\mathbf{x}_i - \mathbf{y}_s) + \epsilon_{is} \tag{6.7}$$

Several authors (Lingoes, 1966; Young, 1972) point out that the unfolding model becomes a simple scaling model if we form a $(n+m) \times (n+m)$ super-matrix from the rectangular distance matrix of the form

$$\begin{array}{c|c} \mathbf{D}_{11} & \mathbf{D}_{12} \\ \hline \mathbf{D}_{21} & \mathbf{D}_{22} \end{array} \tag{6.8}$$

where D_{12} and D_{21} are the rectangular matrix and its transpose, representing the dissimilarities between subjects and stimuli. The diagonal, square submatrices D_{11} and D_{22}, represent the intersubject and interstimulus dissimilarities respectively and simply consist of missing values. Any scaling program that can handle missing data by restricting the summation in the stress formula to non-missing dissimilarities, can therefore be made to fit an unfolding model by construction of this supermatrix. One problem with fitting unfolding models is that the model is not very well constrained due to the missing values on the diagonal submatrices. This is especially true of non-metric row-conditional analysis because here monotonicity is only required to hold within rows, not between rows. As a result, the algorithms are particularly subject to local minima and in extreme cases may lead to degenerate solutions where all subjects or all stimuli are located on a single point. This problem can be overcome by increasing the amount of data relative to the number of points, for example by collecting several replications of the rectangular proximity matrix, for example at different times or for different samples, and minimizing the pooled loss function over all matrices (cf. replicated multidimensional scaling, Section 5.1).

Kruskal (1965) observed that using a modified version of Stress and S-Stress, known as 'formula 2', also makes the solution less prone to local minima and degeneracies than the definitions used in Chapters 3 and 5 (see also Kruskal and Carroll, 1969). Hence the function being minimized is

$$\text{Stress}_2 = \left(\frac{1}{m} \sum_s \frac{\sum_{ij} [\hat{d}_{si} - d_{si}]^2}{\sum (\hat{d}_{si} - \bar{\hat{d}})^2} \right)^{1/2} \tag{6.9}$$

or

$$\text{S-Stress}_2 = \left(\frac{1}{m} \sum_s \frac{\sum_{ij} [(\hat{d}_{si})^2 - (d_{si})^2]^2}{\sum (\hat{d}_{si}^2 - \bar{\hat{d}^2})^2} \right)^{1/2} \tag{6.10}$$

where $\bar{\hat{d}^2}$ and $\bar{\hat{d}}$ are the mean squared disparity and mean disparity respectively. Alternative definitions of S-Stress$_2$ and Stress$_2$ use distances in the denominators instead of disparities.

Greenacre and Browne (1973) developed an alternating least squares algorithm specifically for metric unfolding which minimizes the (unnormalized) sum of squared residuals of (6.7). Here X is estimated keeping Y fixed, then Y is estimated keeping X fixed and this is repeated until convergence.

Example 6.5

As an example of unidimensional scaling, we will look at the famous data from an unpublished PhD thesis by Marks (1965) on the absorption of light at nine different wavelengths by 11 different cones (receptors) in the retina of a goldfish. The data are given in Table 6.5 with the rows and columns permuted as in Hubert and Arabie (1995) to reveal the structure of the matrix (see later).

In the metric unidimensional unfolding solution shown in Fig. 6.5, the colours are ordered more or less according to their wavelengths; the cones are closest to the wavelength that they absorb best.

Table 6.5 Absorption of light of different colours by different cones in goldfish retina (values greater than 100 are underlined).

	Colour (wavelength)								
Cone	Blue-indigo (458 nm)	Violet (430 nm)	Blue (485 nm)	Blue-green (498 nm)	Green (530 nm)	Green (540 nm)	Yellow (585 nm)	Orange (610 nm)	Red (660 nm)
3	153	147	89	57	12	4	0	0	0
8	152	145	125	100	14	0	0	0	0
9	154	153	110	75	32	24	23	17	0
2	101	99	122	140	154	133	93	44	0
6	78	73	85	121	151	154	109	57	0
4	85	46	103	127	152	148	116	75	26
1	2	14	46	52	97	106	137	92	45
7	65	44	77	73	84	102	151	154	120
5	59	87	58	52	86	79	139	153	146
11	27	60	23	24	56	72	136	144	111
10	0	0	40	39	55	62	120	147	132

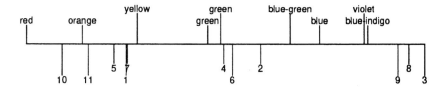

Figure 6.5 Metric unidimensional unfolding of colour vision data (matrix conditional).

The reason this model fits so well is because of the nature of the problem. The colours differ according to a single physical property, wavelength, and the receptor cones have evolved to be able to 'tune in' to different ranges of wavelengths, having maximum sensitivity somewhere in the range. Hence the columns (colours) of the data matrix can be ordered in such a way that, apart from measurement error, each row has a single maximum. Similarly, the rows can be permuted so that for each wavelength, the absorption increases to a maximum down the column of cones and decreases again. Hubert and Arabie (1995) have determined the row and column permutation that gives a matrix which is as close as possible to being a perfectly single peaked matrix (in rows and columns) and this permutation is shown in Table 6.5. Note how similar the ordering is to that in Fig. 6.5. If the data are dichotomized by applying a threshold, 100 say, (underlined numbers in the table) then the fitted matrix will have contiguous ones in its rows and columns. The data may then be represented by displaying a linear scale for the colours and representing each cone by an interval covering the colours which it absorbs at the threshold. □

Permutation of the rows and columns of a dichotomous matrix in order to obtain rows and columns in which the ones are approximately contiguous (i.e. a Petrie matrix) is called *parallelogram analysis*. Dichotomous rectangular data arise in a variety of ways. A market research example is when people are asked to choose a subset of products which appeal to them. The subject by product matrix for such a 'pick any/N' experiment has ones where a subject has chosen a product and zeros elsewhere (see for example Coombs and Smith, 1973). An archaeological example is where the row objects are graves and the columns are different types of objects found in the grave, a '1' indicating that the type of object was found in a grave. Archaeologists would apply parallelogram analysis in order to sort the graves and typology of objects chronologically (i.e. to solve the problem of seriation). The beginning of a column of ones indicates that graves from that time onwards tended to contain that type of object. The ordering of types of objects indicates the order in which different types came into fashion. Kendall (1970) suggests a method for deriving the row and column permutation; simply form a similarity matrix for graves consisting of the number of object types each pair of graves has in common and analyse this matrix by one of the multidimensional scaling methods of Chapter 3. As we showed in Chapter 4, this will usually result in a 'horseshoe' pattern of object types arranged in chronological order.

Dimension 2

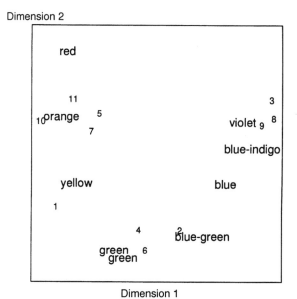

Dimension 1

Figure 6.6   Two-dimensional non-metric unfolding of colour vision data.

Example 6.6

We can also fit a two-dimensional non-metric unfolding solution to the colour vision data. In Fig. 6.6, the colours are clearly arranged in a horseshoe shape and appear in the correct order if one follows the horseshoe from the upper left.   □

Example 6.7

As another example, consider the set of data in Table 6.6. The rectangular matrix consists of the percentages of subjects agreeing that the eight breakfast cereals (columns) have 11 properties (rows). For example, of the eight cereals, the cereal which has the greatest percentage of people agreeing that it is 'fun for children to eat' is Rice Krispies. These rankings can be graphically displayed using multidimensional unfolding. If we are not interested in comparing percentages between rows (e.g. the percentage agreeing that Corn Flakes taste nice with the percentage agreeing that Shredded Wheat stays crispy in milk), then the proper analysis is row conditional. The non-metric row conditional unfolding solution is given in Fig. 6.7.

Any properties which are close together in this figure have similar rankings of cereals. It is therefore interesting to note that 'come back to' and 'reasonably priced' are close together and that 'helps to keep fit' and 'fun for children to eat' are far apart. However, row conditionally implies that the configuration of the cereals should only be interpreted relative to the properties. For example, Corn Flakes is the closest cereal to 'natural flavour', followed by Weetabix and Shredded Wheat. The S-Stress$_2$ value of this solution is 0.06 with the largest contribution from the property 'stays crispy in milk'. Weetabix is too close to

Table 6.6 Percentage of subjects agreeing that breakfast cereals have various properties. The abbreviations used for the properties are defined in Table 6.7.

Properties	Corn Flakes	Weetabix	Rice Krispies	Shredded Wheat	Sugar Puffs	Special K	Frosties	All Bran
CB	65	31	10	10	5	6	6	7
TN	64	40	32	23	29	17	22	11
PF	59	30	20	13	15	7	13	5
ED	60	42	24	18	20	20	19	17
N	40	50	17	31	18	19	14	19
NF	47	39	11	28	6	15	5	18
RP	60	37	9	12	5	6	5	6
FV	27	38	9	26	9	17	7	17
SC	42	6	23	18	13	8	11	6
KF	24	28	10	21	9	29	9	40
FC	17	12	57	5	50	5	43	0

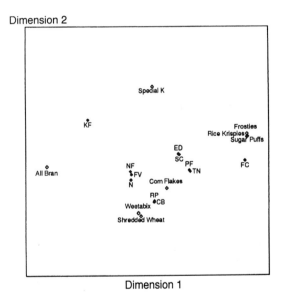

Figure 6.7 Internal unfolding of cereal data.

this property (it should be the furthest away together with All Bran) and Shredded Wheat is too far away (it should be the third closest cereal). □

Vector model

6.3.2 Tucker (1960) suggested another approach to rectangular matrices where both subjects and stimuli are represented by vectors pointing from a common origin so that the relationship between subjects and stimuli is given by

Table 6.7 Key to abbreviations used in Table 6.6 and Figs 6.7 and 6.8.

Abbreviation	Property
CB	come back to
TN	tastes nice
PF	popular with all the family
ED	very easy to digest
N	nourishing
NF	natural flavour
RP	reasonably priced
FV	a lot of food value
SC	stays crispy in milk
KF	helps to keep fit
FC	fun for children to eat

the projection of the stimulus vector, on to the subject vector, or simply by the scalar product of the stimulus and (unit) subject vectors. This model is called the *vector model*.

The vector model is defined analogously to the unfolding model in (6.7) with the distances replaced by the scalar products. It is given by

$$\hat{d}_{si} = \mathbf{x}'_i \mathbf{y}_s + \epsilon_{si} \tag{6.11}$$

The metric solution to the vector model is simply the least squares solution to

$$\mathbf{D} = \mathbf{YX}' + \mathbf{E} \tag{6.12}$$

Let the singular value decomposition be given by $\hat{\mathbf{D}} = \mathbf{USV}'$. Then the least squares approximation to $\mathbf{D}$ in q dimensions is given by $\mathbf{Y} = \mathbf{U}_q \mathbf{S}_q$ (row objects) and $\mathbf{X} = \mathbf{V}_q$ (column objects), where the subscript q denotes the first q rows. The percentage of variation in the disparities explained by the model is given by

$$\frac{\sum_{i=1}^{q} s_i^2}{\sum_{i=1}^{n} s_i^2} \tag{6.13}$$

In a row-conditional analysis, the rows of $\mathbf{Y}$ may be normalized to unit length, since we are only interested in comparing scalar products within subjects, making the relative scaling between subjects irrelevant. The rows of $\mathbf{Y}$ are usually displayed as vectors from the origin whereas the rows of $\mathbf{X}$ are represented by points for the stimuli.

Note that the display for a vector model is similar to a biplot except that in a biplot the rows of the data matrix (samples) are usually represented as points, $\mathbf{Y} = \mathbf{U}_q \mathbf{S}_q$ and the columns (variables) as vectors $\mathbf{X} = \mathbf{V}_q$. Although various different scalings of the vectors $\mathbf{X}$ are in use for biplots, these are only legitimate (unlike the vector model) if $\mathbf{Y}$ is rescaled appropriately so that $\mathbf{Y}_q \mathbf{X}'_q$ approximates the original data matrix. In fact, Gower and Hand (1996) point out the importance of displaying scales on the vectors $\mathbf{X}$ to enable either prediction of the original variable values for each point or, alternatively, interpolation of new data points onto the plot. Note that the vector model in (6.12) becomes a factor analysis model if $\mathbf{D}$ is replaced by a square symmetric (correlation) matrix and $\mathbf{Y}$

is replaced by **X**; factor analysis is simply a vector model for square symmetric proximity data.

Example 6.8

The (metric) vector model is illustrated for the cereal data in Fig. 6.8(a). The two-dimensional solution accounts for 93.8% of the variation in the data. The correlation between the ranked data and the ranked scalar products of the vector model (i.e. the Spearman rank correlation) is 0.84 which is not very different from the rank correlation of 0.89 between the data and the distances of the unfolding model.

It is immediately apparent that 'fun for children to eat' is quite different from the other properties. Figure 6.8(b) shows the projections of the stimulus vectors onto the axes 'fun for children to eat' and 'helps to keep fit'. Here it can be seen that while Rice Krispies, Sugar Puffs and Frosties are 'fun for children to eat', Corn Flakes and Weetabix score highly on all the other properties. The fact that Special K, Shredded Wheat and All Bran are close to the origin implies that their score is low for all the properties. This is not accurate for the property 'helps to keep fit' however, where All Bran has the highest score but only the fourth largest projection. □

Comparison of unfolding and vector models for preference data

6.3.3   In order to compare the unfolding and vector models for preference data, we will discuss an example by Carroll (1972) where the stimuli are cups of tea with different amounts of sugar and at different temperatures. The assumption of the distance model is that the dimensions of the stimulus configuration reflect the perception of the stimuli shared by all subjects, i.e. all subjects differentiate the cups of tea on the basis of their sugar content and temperature. Each subject is represented by a point y_s within the stimulus configuration. The stimulus (cup of tea) closest to this point is the most preferred and the preference for the stimuli decreases with the distance away from the point; the subject point is at the maximum of the subject's *preference function* and is also called the subject's *ideal point*. This ideal point corresponds to the most preferred combination of attributes associated with the dimensions, i.e. the ideal sugar content and temperature.

In the vector model, the preference function for a given subject increases as the projection of the stimulus vectors onto the subject vector increases. Since this projection depends linearly on the coordinates of the stimuli, the preference function changes linearly and therefore monotonically with the dimensions. The interpretation in the tea example is that if a subject likes some sugar, he or she will like more sugar even better. There is no ideal amount of sugar and no amount of sugar will ever be too much. This kind of monotonically increasing preference function is rarely appropriate psychologically and the distance model is therefore often considered more acceptable than the vector model. Another advantage of the distance model is that it includes the vector model as special case.

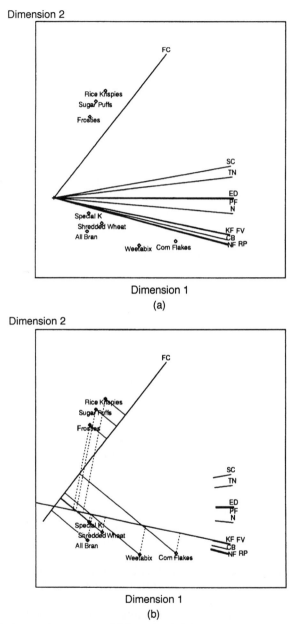

Figure 6.8   Internal vector model for cereal data.

This can be seen by considering the preference function and its contours, the *isopreference contours*. In the vector model, all stimuli with the same projection onto the subject vector are equally preferred and therefore the isopreference contours are a set of parallel lines (planes in three dimensions) perpendicular to the subject vector. In the distance model, all stimuli equidistant from an ideal point are equally preferred and therefore the isopreference contours are a set of concentric circles (or spheres in three dimensions) centred on the ideal point. When an ideal point moves further and further away from the origin, the circular isopreference contours in the part of the space occupied by the stimuli look more and more like parallel straight lines perpendicular to the line joining the origin and the ideal point—the isopreference contours of the vector model.

Carroll's hierarchy of models

6.3.4   Carroll's hierarchy of models for rectangular matrices (Carroll, 1972) starts with the vector model at the lowest level and the distance model one level up. The distance model is further generalized by allowing the dimensions to be weighted differently for each subject. The weighted unfolding model is given by

$$d_{sr}^2 = (\mathbf{x}_i - \mathbf{y}_s)'\mathbf{W}_s(\mathbf{x}_i - \mathbf{y}_s) \tag{6.14}$$

where $\mathbf{W}_s$ is a diagonal matrix of weights. This model allows some dimensions to be more important than others in determining a subject's preferences; the isopreference contours become ellipsoids. Finally, in the general unfolding model, the requirement that the weight matrix $\mathbf{W}_s$ is diagonal is relaxed so that each subject has his or her own set of dimensions along which the space is weighted; the principal axes of the isopreference ellipsoids are now allowed to make different angles with the coordinate axes.

Carroll notes that a negative weight along one of the dimensions indicates the least preferred value along that dimension. This will be more appropriate than an ideal point when the preference function has a minimum. For example, most people like hot tea and iced tea but dislike tepid tea. These people's least preferred point would be at some intermediate temperature. The weighted unfolding model and general unfolding model are difficult to fit in practice except in an *external analysis*.

External analyses

6.3.5   In some instances, we may have, in addition to a rectangular matrix relating subjects to stimuli, some further knowledge about the interrelationship between the stimuli. For example, the stimuli may be geographic locations or we may already have derived a multidimensional scaling solution from a dissimilarity matrix of the stimuli. We may wish simply to add subjects to the existing configuration of stimuli so that the distances (or scalar products) between subjects and stimuli match as closely as possible the elements of the rectangular matrix. This type of analysis is called an *external* analysis because the subject's preferences can be viewed as external information added to the known configurations of stimuli. External analyses also provide a useful method for avoiding local minima and degeneracies frequently encountered in unfolding. In fact, methods have been proposed for external unfolding analysis in the absence of additional data on pairwise proximities of stimuli, by deriving the stimulus configuration directly from the rectangular matrix (Rabinowitz, 1976; Rodgers and Young, 1981).

For the external vector model it is the scalar product of the external vector $\mathbf{y}_s$ and the stimulus position vector $\mathbf{x}_i$ that should be linearly related to the dissimilarity δ_{si}, i.e. $\delta_{si} = a_s \mathbf{x}'_i \mathbf{y}_s + c_s + \epsilon_{si}$. If the subject vector $\mathbf{y}_s$ is of unit length and has direction cosines b_{jk} in the coordinate system of $\mathbf{X}$, (satisfying $\sum_k b_{jk}^2 = 1$), then the scalar product $\mathbf{x}'_i \mathbf{y}_s$ is equal to $\sum_{k=1}^{q} b_{sk} x_{ik}$. The external vector model can therefore be written as

$$\delta_{si} = a_s \mathbf{x}'_i \mathbf{y}_s + c_s + \epsilon_{si}$$

$$= a_s \sum_{k=1}^{q} b_{sk} x_{ik} + c_s + \epsilon_{si}$$

$$= \sum_{k=1}^{q} \beta_{sk} x_{ik} + \alpha_s + \epsilon_{si} \tag{6.15}$$

Thus, regressing δ_{si} on x_{ik} gives the regression parameters $\hat{\beta}_{sk} = a_s b_{sk}$ from which the direction cosines of $\mathbf{y}_s$ can be derived by normalization

$$b_{sk} = \hat{\beta}_{sk} \Big/ \left(\sum_k \hat{\beta}_{sk}^2 \right)^{1/2}$$

This method has already been discussed in Chapter 3 in the context of interpreting multidimensional scaling solutions.

For the external distance model, it is the squared distance between the external point $\mathbf{y}_s$ and the stimulus point $\mathbf{x}_i$ that should be linearly related to δ_{si}. If $\mathbf{y}_s$ has coordinates b_{sk} along the dimensions of $\mathbf{X}$, we have

$$\delta_{si} = a_s (\mathbf{x}_i - \mathbf{y}_s)'(\mathbf{x}_i - \mathbf{y}_s) + c_s + \epsilon_{si}$$

$$= a_s \left(\sum_k x_{ik}^2 - \sum_k 2 b_{sk} x_{ik} + \sum_k b_{sk}^2 \right) + c_s + \epsilon_{si}$$

$$= \gamma_s \sum_k x_{ik}^2 + \sum_k \beta_{sk} x_{ik} + \alpha_s + \epsilon_{si} \tag{6.16}$$

Linear least squares yields estimates, $\hat{\gamma}_1 = a_s$ and $\hat{\beta}_{sk} = -2 a_s b_{sk}$ and $\hat{\alpha}_s = c_s + a_s \sum_k b_{sk}^2$ so that the coordinates of $\mathbf{y}_s$ are given by $b_{sk} = -\hat{\beta}_{sk}/2\hat{\gamma}_s$.

It can be seen from equations (6.15) and (6.16) that the external vector model is a special case of the external distance model so that the usual F-statistics can be used to decide which model is more appropriate. Further, these models are nested within the weighted model, which in turn is nested within the general unfolding model. The latter two models are solved using similar regressions to (6.16) followed by the estimation of the weight matrices $\mathbf{W}_s$ from the regression parameters, see Carroll (1972) for details. We can select from the complete hierarchy of models simply by comparing F-statistics.

For a non-metric analysis, the left-hand side in equations (6.15) and (6.16) is replaced by $\hat{d}_{si} = f^{(s)}(\delta_{si})$ where $f^{(s)}$ is a monotonic function. The algorithm works as follows. First estimate the regression equation in (6.15) or (6.16). Then apply least squares monotone regression to find the first estimate of the monotonic function (see Chapter 3) and hence of $\hat{d}_{si}$. Use these first estimates of the disparities as the new response variable in the regression. Keep updating the disparities and coefficients until convergence is achieved.

Example 6.9

Delbeke (1968) asked a group of 80 psychology students at the University of Louvain to make preference judgements on all possible pairs of 20 family compositions, as described by the numbers of sons and daughters. We will fit Carroll's hierarchy of models to a subset of these data (18 subjects and 18 family compositions) following Coxon (1982a). The first three columns of Table 6.9 show the data for the first subject. The stimulus configuration can simply be defined as a square lattice of points with dimension 1 equal to the number of boys and dimension 2 the number of girls in the family. The objective is to place the subjects into the stimulus space by selecting from the vector, distance, weighted and general unfolding models using F-ratios as a guide. Table 6.8 summarizes the correlations and F-statistics obtained for the four models using a metric analysis.

It is clear that the correlations increase substantially from the vector model to the distance model but do not increase much further for the weighted and general distance models. The fit of all models except the vector model is highly significant for most subjects (see p_w in Table 6.8). The only model that provides a significant improvement over the model one level down in the hierarchy is the simple distance model (see p_b in Table 6.8). All these facts seem to suggest that the simple distance model is the best model. The subject configuration within the stimulus space defined by the dimensions 'boys' and 'girls' is shown in Fig. 6.9.

It is apparent that most of the subjects prefer large families with almost as many girls as boys. Table 6.9 shows the data for subject one, together with the disparities (preferences standardized to mean zero and variance 100 and reversed in sign), the squared distances (between the subject point and the stimulus points), and the best-fitting linearly transformed squared distances and residuals.　　　　　　　　　　　　　　　　　　　　　　　　　　　□

Table 6.8 Correlations and F-statistics for family composition data. *F*-ratios and *p*-values are for testing the significance of the model (*w*) and selecting between the model and the model one level down in the hierarchy (*b*).

Model (no. of parameters)	Vector (3) min	max	Distance (4) min	max	Weighted (5) min	max	General (6) min	max
r	0.15	0.65	0.70	0.93	0.70	0.94	0.70	0.94
r (root mean square)	0.47		0.83		0.84		0.84	
F_w	0.2	5.4	4.5	32.3	3.1	22.8	2.4	19.2
p_w	0.83	0.02	<0.001	0.02	<0.001	0.05	<0.001	0.1
F_b			11.7	69.2	0.0	1.1	0.0	1.5
p_b			0.004	<0.001	0.99	0.31	0.94	0.24

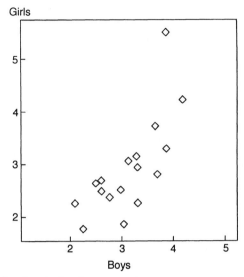

Figure 6.9   Ideal points in family composition space.

Table 6.9   Distance model for subject 1 with ideal point coordinates (2.77, 2.37).

Boys	Girls	Preferences	Squared distances	Transformed Squared distances	Disparities	Residuals
1	0	6	8.75	14.22	18.75	−4.53
2	0	12	6.21	0.08	−7.21	7.29
3	0	10	5.68	−2.92	1.44	−4.36
4	0	7	7.14	5.22	14.42	−9.20
5	0	5	10.69	24.50	23.08	1.42
0	1	6	9.55	18.66	18.75	−0.09
0	2	8	7.81	8.96	10.10	−1.14
0	3	4	8.07	10.39	27.41	−17.02
0	4	3	10.32	22.97	31.73	−8.76
0	5	1	14.58	46.69	40.39	6.30
1	1	16	5.01	−6.62	−24.52	17.90
4	1	17	3.39	−15.62	−28.85	13.23
1	2	14	3.27	−16.32	−15.87	−0.45
2	2	13	0.73	−30.46	−11.54	−18.92
3	2	16	0.19	−33.47	−24.52	−8.95
1	3	20	3.53	−14.89	−41.83	26.94
2	3	17	0.99	−29.03	−28.85	−0.18
1	4	11	5.78	−2.31	−2.88	0.57

Three-way asymmetric and rectangular data

6.4 So far we have discussed the situation of several symmetric matrices (Chapter 5) and of an individual asymmetric or rectangular matrix (current chapter). How would we analyse a collection of asymmetric or rectangular matrices? This can be done using the general Euclidean model (Young, 1987). Indexing the matrices by o for occasions $o = 1, 2, \ldots, l$, the general Euclidean model is given by

$$d_{ij}^{(s)} = (\mathbf{x}_i - \mathbf{y}_s)'\mathbf{V}_s\mathbf{W}_o(\mathbf{x}_i - \mathbf{y}_s)$$

where $\mathbf{V}_s$ and $\mathbf{W}_o$ are $(q \times q)$ positive semi-definite matrices. The matrix $\mathbf{W}_o$ contains the squared dimension weights, similar to those of the weighted Euclidean model in Chapter 5 and $\mathbf{V}_s$ are the row weights of the model in Section 6.3. The general Euclidean model includes many of the models discussed in Chapters 3 and 5 as special cases. For example, the scaling model for a two-way symmetric matrix is obtained by setting $\mathbf{W}_o = 1$, $\mathbf{V}_s = 1$ and $\mathbf{x}_i = \mathbf{y}_i$. Applying only some of these constraints yields many of the more complex models.

For three-way asymmetric dissimilarity matrices, we have $\mathbf{V}_s = 1$ and $\mathbf{x}_i = \mathbf{y}_i$ and for three-way rectangular matrices, $\mathbf{V}_s = 1$ and $\mathbf{W}_o$ is usually taken to be diagonal.

Summary

6.5 A number of methods are available for scaling square asymmetric data. Perhaps the most useful for visualizing the amount of asymmetry between pairs of stimuli is the canonical decomposition of skew symmetric matrices.

Rectangular matrices, as well as square asymmetric matrices, may be represented by two sets of points, one set for the row objects and one set for the column objects. In unfolding models, the distances between these sets of points approximate the elements of the rectangular matrix. In vector models, the scalar products between the position vectors of the two sets of points approximate the elements of the rectangular matrix. Both these models can also be fitted externally, by simply adding the row information to an existing configuration of the column points. The general Euclidean model combines the models for three-way data described in Chapter 5 with the models for square asymmetric and rectangular data described in this chapter to deal with more complex data.

7

Tree Models for Proximity Data

Introduction

7.1   The spatial models described in previous chapters represent each object or stimulus of interest as a point in a coordinate space so that metric distances between the points reflect the relationships between the stimuli as implied by the observed proximities. An alternative approach in which stimuli are represented by nodes in a connected graph without cycles, i.e. a *tree* (see Chapter 4), is discussed in this chapter. The terminal nodes in such a structure represent the stimuli, and each pair of nodes is joined by a unique path so that the relationship between them approximates in some sense the observed similarities or dissimilarities.

A tree structure lends itself to a natural interpretation as a *hierarchical clustering scheme* (see Johnson, 1967) and the hierarchical or *nested* structures implied by trees often provide good descriptions of the proximity data; in particular they are generally easy to interpret and frequently convey important information about the domain of the stimuli.

Before describing the various tree models useful for proximity data, some graph-theoretical concepts used later are briefly defined. (The material in this chapter owes much to the excellent introduction to tree models given in Corter, 1996.)

Some graph-theoretical concepts

7.2   A graph (V, E) consists of a set of nodes, $V = \{v_1, \ldots, v_n\}$ and a set of links or edges, $E = \{(v_i, v_j) | v_i, v_j \in V\}$ connecting two nodes. In a *directed* graph elements in E are ordered pairs; in an *undirected* graph the pairs (v_i, v_j) are unordered. The *degree* of a node v_i is defined as the number of links incident in v_i. Nodes with degree one are *terminal nodes*, while nodes with a degree larger than one are non-terminal or *internal nodes*. A *path* between two nodes v_a and v_z is a sequence of connected links $(v_a, v_b), (v_b, v_c), \ldots, (v_x, v_y), (v_y, v_z)$. A graph is *connected* if there exists a path between any two elements of V. In a *weighted graph*, a non-negative weight, $w(v_i, v_j)$ is associated with each link in E. The *length* of a path is then defined as the sum of the weights assigned to the links that occur in the path; i.e. the length of the path from v_a to v_z is $w(v_a, v_b) + w(v_b, v_c) + \ldots + w(v_y, v_z)$. The *minimum path length distance* between any two nodes $v_i, v_j \in V$, is defined as the minimum length of the paths that connect v_i and v_j. A *tree* is a connected graph where every pair of nodes is connected by a *unique* path. This restriction implies that in a tree there are no links (v_i, v_i) that connect a node v_i to itself.

Additive and ultrametric trees

7.3   The two types of tree model that have most often been applied to proximity data are *additive* trees and *ultrametric* trees. We begin with an account of the former.

Additive trees

7.3.1   An additive tree is a connected, undirected graph where every pair of nodes is connected by a unique path. Since there exists only one path between any two nodes, the minimum path length distance between two nodes is equal to the length of the unique path that connects them. These distances are often referred to as *path length distances* and additive trees are sometimes called *path length trees* (Carroll, 1976; Carroll and Chang, 1973). An example of an additive tree appears in Fig. 7.1.

The defining feature of an additive tree is a relationship between the path length distances known as the *additive inequality* or the *four-point condition* (see Buneman, 1971; Dobson, 1974; Patrinos and Hakimi, 1972). Consider, for example, the additive tree shown in Fig. 7.2, with path lengths as indicated.

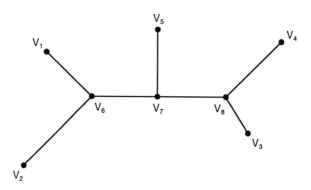

Figure 7.1   Additive tree. (Reproduced with permission from De Soete and Carroll, 1996.)

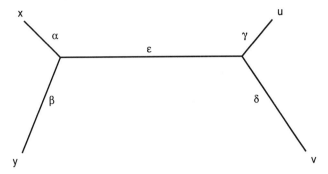

Figure 7.2 Additive tree showing path lengths.

Letting d_{ij} represent the additive tree distance between the nodes i and j, it follows in this case that

$$d_{x,y} + d_{u,v} = \alpha + \beta + \gamma + \delta \qquad (7.1)$$

$$\leqslant \alpha + \beta + \gamma + \delta + 2\epsilon \qquad (7.2)$$

$$= d_{x,u} + d_{y,v} \qquad (7.3)$$

$$= d_{x,v} + d_{y,u} \qquad (7.4)$$

Any four nodes in a set of interest can be labelled so as to satisfy the above inequality; consequently in an additive tree

$$d_{x,y} + d_{u,v} \leqslant \max[d_{x,u} + d_{y,v}, d_{x,v} + d_{y,u}] \text{ for all } x, y, u \text{ and } v \qquad (7.5)$$

The additive inequality is both necessary and sufficient for a set of distances to define a unique additive tree (for a proof see Buneman, 1971, or Dobson, 1974). Consequently, if a set of observed dissimilarities satisfies the additive inequality, the dissimilarities can be represented perfectly by the corresponding additive tree. Corter (1996), for example, gives the following dissimilarity matrix, in which it is not hard to verify that the dissimilarities among every quadruple satisfy the additive inequality

$$
\Delta = \begin{array}{c} \\ A \\ B \\ C \\ D \\ E \end{array}
\begin{array}{c}
\begin{array}{ccccc} A & B & C & D & E \end{array} \\
\left(\begin{array}{ccccc}
- & & & & \\
15 & - & & & \\
20 & 25 & - & & \\
18 & 23 & 6 & - & \\
20 & 25 & 20 & 18 & -
\end{array}\right)
\end{array}
\qquad (7.6)
$$

The corresponding additive tree representation is shown in Fig. 7.3.

In this artificially constructed example there is a perfect correlation between the observed dissimilarities and the path lengths in the additive tree. Most empirically derived data will, however, not exactly satisfy the additive inequality for every quadruple of dissimilarities. The consequence is that such data cannot be represented perfectly by an additive tree. The goal now becomes that of

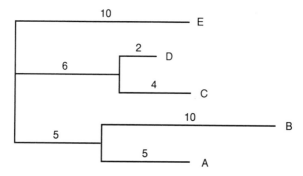

Figure 7.3 Additive tree representation of dissimilarity matrix satisfying the additive inequality. (Reproduced with permission from Corter, 1996.)

finding an additive tree structure that models the proximity data as accurately as possible in the sense that the path distances in the tree are as close as possible (as judged say, by a least squares criterion) to the original dissimilarities, i.e. a tree for which SSE given by

$$\text{SSE} = \sum (d_{ij} - \delta_{ij})^2 \tag{7.7}$$

is minimized.

Algorithms for fitting additive trees to proximity data are described in Section 7.4.

Ultrametric trees

7.3.2 An ultrametric tree is an additive tree in which each terminal node is equidistant from some specific node, called the *root* of the tree. Except for the root, the degree of all internal nodes is at least three. An example of an ultrametric tree is shown in Fig. 7.4; in this tree v_9 is the root.

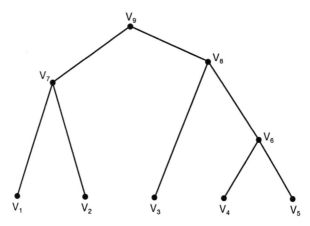

Figure 7.4 Ultrametric tree. (Reproduced with permission from De Soete and Carroll, 1996.)

The requirement that the distances from the terminal nodes to the root are equal implies a series of constraints on the path distances in an ultrametric tree; in the tree in Fig. 7.4, for example, the requirement leads to the following equalities

$$d_{17} = d_{27} \tag{7.8}$$

$$d_{46} = d_{56} \tag{7.9}$$

$$d_{38} = d_{46} + d_{68} \tag{7.10}$$

$$d_{17} + d_{79} = d_{38} + d_{89} \tag{7.11}$$

(See De Soete and Carroll, 1996, for details).

Because of the constraints implied by the equidistance condition, the path lengths in an ultrametric tree can be represented more parsimoniously in terms of numerical values associated with each node (sometimes referred to as the heights of the nodes), rather than with the edges. These values are such that:

(1) for the terminal nodes the values are zero;
(2) the root has the largest value;
(3) the values attached to the nodes on the path from any terminal node to the root constitute an increasing sequence.

The distance between any two terminal nodes is now equal to the largest height value attached to any of the nodes that occur on the path connecting the two nodes.

Ultrametric trees in which the nodes represent objects or stimuli of interest are familiar from the field of cluster analysis (see Everitt, 1993), and such a tree defines a hierarchical clustering of the set of objects. In Fig. 7.4, for example, if the following is assumed for the heights (h) associated with each node

$$0 < h(v_6) < h(v_7) < h(v_8) < h(v_9) \tag{7.12}$$

Then each level of the tree defines a partition of the five objects. At the lowest level, each object constitutes its own subset in the partition: {1}, {2}, {3}, {4}, {5}. At the next level (v_6) the partition becomes {1}, {2}, {3}, {4, 5}, followed at v_7 by {1, 2}, {3}, {4, 5}, and at v_8 by {1, 2}, {3, 4, 5}. Finally at the highest level, the root v_9, we have a single set comprising all the objects, {1, 2, 3, 4, 5}.

In terms of such a series of partitions, the distance between two terminal nodes i and j in an ultrametric tree can also be defined as the height corresponding to the smallest subset containing both i and j. These distances, which we shall again represent as d_{ij}, can be shown to satisfy the ultrametric inequality (Johnson, 1967)

$$d_{ij} \leqslant \max[d_{ik}, d_{jk}] \tag{7.13}$$

Equivalently, for all triples (i, j, k), the two largest values in the set $\{d_{ij}, d_{ik}, d_{jk}\}$ are equal so that

$$d_{ij} \leqslant d_{ik} = d_{jk} \tag{7.14}$$

If a set of observed dissimilarities satisfies the ultrametric inequality, then the dissimilarities can be represented exactly by a tree model. Corter (1996), for

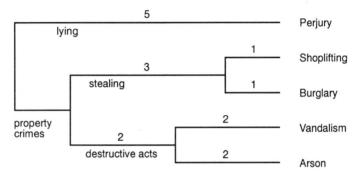

Figure 7.5   Ultrametric tree representation of dissimilarity judgements of crimes.

example, gives the following matrix representing hypothetical judgements of dissimilarity among five crimes: 1. arson, 2. burglary, 3. perjury, 4. shoplifting and 5. vandalism

$$\Delta = \begin{array}{c} \\ 1 \\ 2 \\ 3 \\ 4 \\ 5 \end{array} \begin{array}{ccccc} 1 & 2 & 3 & 4 & 5 \\ \left(\begin{array}{ccccc} - & & & & \\ 8 & - & & & \\ 10 & 10 & - & & \\ 8 & 2 & 10 & - & \\ 4 & 8 & 10 & 8 & - \end{array} \right) \end{array} \qquad (7.15)$$

Figure 7.5 shows the ultrametric tree representation of Δ. The tree organizes highly similar crimes (e.g. burglary and shoplifting) into the same branch of the tree.

If a set of observed dissimilarities satisfies the ultrametric inequality perfectly, the dissimilarities can be uniquely represented by an ultrametric tree whose inter-object distances are exactly equal to the corresponding dissimilarities. But as with the additive trees described earlier, it is rare for empirical proximity data to satisfy the required ultrametric condition perfectly, and so it becomes necessary to consider how to arrive at a 'best' fitting tree. The problem will be considered in Section 7.4.

Additive and ultrametric trees compared

7.3.3   Additive trees and ultrametric trees are defined explicitly by the additive and ultrametric inequalities on path lengths given in (7.5) and (7.13). The additive inequality is, in fact, implied by the ultrametric inequality, and it is easy to see that the former is less restrictive than the latter by considering any four points on a line; the interpoint distances satisfy the former but not the latter. The greater flexibility of the additive tree arises from the number of parameters involved compared with an ultrametric tree; in the latter all $n(n-1)/2$ inter-object distances are determined by the heights of the $n-1$ nodes in the tree. In an additive tree, the distances are determined by at most $2n-3$ parameters (see Sattath and Tversky, 1977).

The most obvious difference between ultrametric trees and additive trees is that, in the former, all terminal nodes are required to be an equal distance from the root node. In an additive tree, path lengths are not so constrained and there is no specific root node. Consequently, such trees can be equally well displayed in *unrooted* (see Fig. 7.1) or *rooted* (see Fig. 7.3) form. It is, however, generally more convenient to display trees in a rooted form, but since an additive tree can be rooted at any point in the tree graph, the question arises as to how to choose the root. Such a choice can greatly affect the interpretation of an additive tree and might be seen as analogous to the choice of rotation in factor analysis. This problem will be returned to later in the chapter.

Carroll and Pruzansky (1975) and Carroll *et al.* (1984) prove that an additive tree can always be decomposed into the sum of an ultrametric tree and a *star tree*, the latter being an additive tree with exactly one internal node. Because of its special structure, the distance $d_{ij}^{(S)}$ between any two terminal nodes i and j of a star is of the form

$$d_{ij}^{(S)} = s_i + s_j \qquad (7.16)$$

where s_i denotes the weight attached to the link connecting terminal node i with the internal node (see Fig. 7.6). The distance between two terminal nodes i and j of an additive tree, $d_{ij}^{(A)}$ can be decomposed into the sum of an ultrametric distance $d_{ij}^{(U)}$ and a distance in a star tree

$$d_{ij}^{(A)} = d_{ij}^{(U)} + d_{ij}^{(S)} = d_{ij}^{(U)} + s_i + s_j \qquad (7.17)$$

The decomposition is not unique (see Brossier, 1985).

Algorithms for fitting trees to data

7.4   Since dissimilarity data collected in practice rarely if ever satisfy perfectly the additive or ultrametric inequalities, a method is needed which constructs

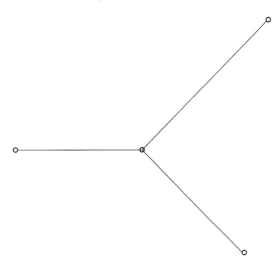

Figure 7.6   Star tree.

either an additive tree or ultrametric tree in which path length distances are as close as possible in some sense to the observed dissimilarities. The fit function most often used is least squares, i.e.

$$L(\mathbf{D}) = \sum_{i<j} w_{ij}(d_{ij} - \delta_{ij})^2 \tag{7.18}$$

where $\mathbf{D}$ is the matrix of tree distances, d_{ij}, and δ_{ij} represents the observed dissimilarity of stimuli i and j. In many applications the weights w_{ij} will be set equal to one, but when dissimilarities are missing, the corresponding weight can be set equal to zero.

Finding either an additive or ultrametric tree to represent a set of observed dissimilarities can be formulated in essentially similar fashion namely:

(1) additive tree

$$\operatorname*{minimize}_{\mathbf{D}} \; L(\mathbf{D})$$

subject to $d_{ij} + d_{kl} \leqslant \max[d_{ik} + d_{jl}, d_{il} + d_{jk}]$ for all i, j, k, l $\qquad$ (7.19)

(2) ultrametric tree

$$\operatorname*{minimize}_{\mathbf{D}} \; L(\mathbf{D})$$

subject to $d_{ij} \leqslant \max[d_{ik}, d_{kj}]$ for all i, j, k $\qquad$ (7.20)

Various authors have attacked this optimization problem, the earliest being Hartigan (1967) who iteratively transforms an existing tree using local operations such as relocating branches and deleting internal nodes. Carroll and Pruzansky (1975) describe a mathematical programming approach to the problem that, under certain conditions, can be guaranteed to find the optimal solution. Perhaps the most practical algorithm is that suggested by De Soete (1984a, b), who showed that the constrained optimization problem stated above can be transformed into a series of unconstrained problems, involving the sequential minimization of the augmented function, $\Phi(\mathbf{D}, \rho)$, given by

$$\Phi(\mathbf{D}, \rho) = L(\mathbf{D}) + \rho P(\mathbf{D}) \; (\rho > 0) \tag{7.21}$$

for an increasing sequence of values of ρ. The function Φ is a linear combination of the fit function L and a penalty function, P, used to enforce either the additive or ultrametric inequality on $\mathbf{D}$. The penalty function for an ultrametric tree is defined as follows

$$P(\mathbf{D}) = \sum_{\Omega_{\mathbf{D}}} (d_{ik} - d_{jk})^2 \tag{7.22}$$

where $\Omega_{\mathbf{D}}$ denotes the set of ordered triples (i, j, k) for which $\mathbf{D}$ violates the ultrametric inequality

$$\Omega_{\mathbf{D}} = \{(i, j, k) | d_{ij} \leqslant \min[d_{ik}, d_{jk}] \text{ and } d_{ik} \neq d_{jk}\} \tag{7.23}$$

For an additive tree the definition of the penalty function is

$$P(\mathbf{D}) = \sum_{\Omega_D} (d_{ik} + d_{jl} - d_{il} - d_{jk})^2 \qquad (7.24)$$

where

$$\Omega_D = \{(i, j, k, l) \mid d_{ij} + d_{kl} \leqslant \min[d_{ik} + d_{jl}, d_{il} + d_{jk}] \text{ and } d_{ik} + d_{jl} \neq d_{il} + d_{jk}\} \qquad (7.25)$$

Using r as the iteration index, the procedure outlined by De Soete (1984a, b) for performing the optimization is as follows.

Step 1. Initialize r: $r = 1$. Determine $\mathbf{D}^{(0)}$ the initial estimate of $\mathbf{D}$ as

$$d_{ij}^{(0)} = \delta_{ij} + \epsilon_{ij} \qquad (7.26)$$

where ϵ_{ij} is a random normal variable with mean zero and variance equal to one third of

$$\frac{2}{n(n-1)} \sum_{i<j} (\delta_{ij} - \bar{\delta})^2 \qquad (7.27)$$

where

$$\bar{\delta} = \frac{2}{n(n-1)} \sum_{i<j} \delta_{ij} \qquad (7.28)$$

Step 2. Define $\rho^{(1)} = L(\mathbf{D}^{(0)})/P(\mathbf{D}^{(0)})$.
Step 3. Minimize $\Phi(\mathbf{D}, \rho^{(r)})$ starting from $\mathbf{D}^{(r-1)}$ to obtain $\mathbf{D}^{(r)}$.
Step 4. Test for convergence: if $\sum_{i<j}(d_{ij}^{(r)} - d_{ij}^{(r-1)})^2$ is less than some small constant stop, otherwise continue.
Step 5. Update ρ; $\rho^{(r+1)} = 10 \times \rho^{(r)}$. Increment r by one and return to Step 3.

In Step 3 an unconstrained minimization problem has to be solved. De Soete (1984a,b) uses the conjugate gradient procedure suggested by Powell (1977).

The algorithm outlined above only guarantees convergence to a local minimum, but in a Monte Carlo evaluation of its performance, De Soete (1984c) finds that the algorithm is relatively insensitive to the choice of initial parameter estimates and that it probably often converges to the global optimum.

Choosing a root for an additive tree

7.5   In an ultrametric tree, there is a unique point in the tree that is equidistant from all terminal nodes; this is the root node. In an additive tree, however, the root is not determined by the distances and any point on the tree can serve as a root. But different roots may suggest different interpretations of the common or distinctive features of the stimuli, since they induce different hierarchies of partitions or clusters. If, for example, the root is placed along any path in the tree, there will, in general, be exactly two clusters distinguished by the root, that is two subtrees that descend from the root. If, however, the root is placed exactly at an internal node (which is usually of degree three), there will be three subtrees that descend from the root, corresponding to three major clusters of stimuli.

The choice of a root is therefore clearly of some importance in using additive tree representations of proximity data in practice. Researchers using additive trees need to be alert to the possibility of alternative rootings that might provide differing perspectives on the data. A number of procedures have been suggested for choosing a root. Corter (1996), for example, proposes choosing a root so as to enhance the interpretability of the tree. This, however, appears to imply prior knowledge of structure which fitting the tree is aimed to recover. A more objective method for choosing a root is described by Sattath and Tversky (1977), who propose placing the root at the point in the tree that minimizes the variance of the distances from the root to the terminal nodes. An interactive procedure for selecting a root that leads to the most interpretable tree has been devised by De Soete and Vermeulen (1993).

7.6   Some examples of trees fitted to proximity data

Example 7.1

De Soete and Carroll (1996) illustrate the least squares ultrametric tree fitting procedure described in Section 7.4 on some data collected by Rosenberg and Kim (1975) concerning the similarity between kinship terms. Subjects were asked to group 15 kinship terms on the basis of their similarities into between 2 and 15 categories. Dissimilarity values for each pair of terms were calculated by counting the number of subjects who placed the two terms in different categories. The derived dissimilarity matrix for female subjects taken from Rosenberg (1982) is shown in Table 7.1. The resulting ultrametric tree found using the algorithm described in the previous sections is shown in Fig. 7.7.

Table 7.1   Dissimilarity matrix for kinship terms. Number of subjects (out of 85) who did not put any two terms into the same category

	1	2	3	4	5	6	7	8	9	10	11	12	13	14	15
1. Grandfather	0														
2. Grandmother	11	0													
3. Granddaughter	48	38	0												
4. Grandson	38	48	13	0											
5. Sister	83	76	73	82	0										
6. Brother	76	83	82	73	10	0									
7. Mother	82	73	74	83	55	63	0								
8. Father	73	82	83	74	63	55	13	0							
9. Daughter	83	74	69	80	52	61	34	43	0						
10. Son	74	83	80	69	61	52	43	34	14	0					
11. Nephew	78	84	80	73	81	74	83	77	81	74	0				
12. Niece	85	79	72	79	76	83	79	85	74	81	12	0			
13. Cousin	84	84	82	81	78	77	84	84	83	82	53	53	0		
14. Aunt	84	78	79	85	77	83	72	79	79	85	49	42	38	0	
15. Uncle	78	84	85	79	83	77	79	72	85	79	42	49	39	10	0

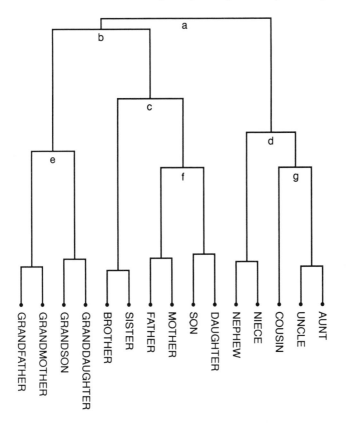

Figure 7.7   Ultrametric tree for kinship data. (Reproduced with permission from De Soete and Carroll, 1996.)

In Fig. 7.7 the root node (labelled a), distinguishes the direct kin (grand-parents, grandchildren, parents, brother, sister) from the collaterals (uncle, aunt, cousin, nephew, niece). Within the former, node b separates the nuclear family from the kin that are two generations away. Node c distinguishes the members of the nuclear family from the same generation (brother, sister) from those one generation apart. The derived ultrametric tree accounts for 96% of the variance amongst the observed dissimilarities; consequently it gives a very accurate representation of these values.   □

Example 7.2

To illustrate the use of an additive tree, dissimilarity ratings between animals, averaged over 18 subjects, as reported by Henley (1969), will be used. Each subject rated the dissimilarity betwen all pairs of 30 animals on a scale from 0 to 10. The results of applying the algorithm for fitting additive trees to these data are presented in Fig. 7.8. The resulting tree is shown in parallel form, i.e. all the branches are parallel, and the distance between two nodes is the sum of the horizontal sections of the path joining them. The root was chosen by minimizing the variance of the distance from the terminal nodes as described in Section 7.5.

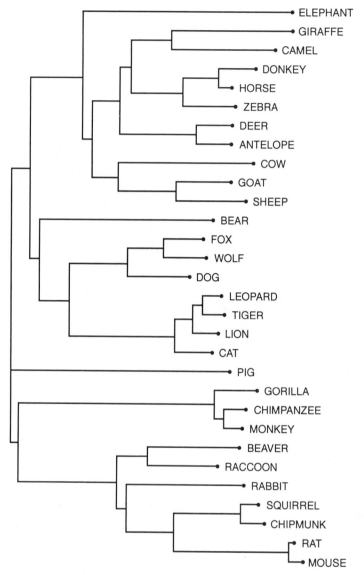

Figure 7.8 Additive tree for animal data. (Reproduced with permission from De Soete and Carroll, 1996.)

In Fig. 7.8 the animals are first partitioned into four major clusters; herbivores, carnivores, apes and rodents. Each of these clusters is then further partitioned into finer clusters. The carnivores, for example, are partitioned into felines (cat, leopard, tiger and lion), canines (dog, fox, wolf) and bear. The vertical order of the animals in Fig. 7.8 corresponds roughly to size, with elephant and mouse at the two endpoints. The (horizontal) distance of an animal from the root reflects its average distance from other animals. Cat, for example, is closer to the root than tiger, because it is more similar, on average to the other animals than tiger. This property of the data could not be represented in an ultrametric tree in which all objects are equidistant from the root. □

Tree representations of more complex proximity data

7.7 Tree models have been proposed for both three-way two-mode and for rectangular proximity data (including asymmetric dissimilarity matrices). Such models do not appear to have been widely used in practice and so only a brief account of the methods that have been suggested will be given here.

Three-way two-mode data

7.7.1 One simple way of fitting a tree model to three-way two-mode data would be simply to average the separate proximity matrices and then fit either an additive or ultrametric tree to the resulting single proximity matrix using one of the methods described in Section 7.5. Such an approach would, as previously mentioned in Chapter 5, only be of use if differences between the individual proximity matrices could be assumed to be due largely to 'error'. The averaging procedure would, however, mask any systematic differences between the individual proximity matrices and thus prevent the investigator from exploring the feature usually of most concern.

An alternative approach to simple averaging is to postulate a model that incorporates the assumption that every individual has the *same* tree structure or topology underlying their judgements, but that they weigh different paths differently (cf. three-way scaling as described in Chapter 5). The requirement that the individual trees have the same topology implies, for an ultrametric tree, that each triple (i, j, k) satisfies the ultrametric inequality in the *same* way; for an additive tree the corresponding constraint is that each quadruple (i, j, k, l) satisfies the additive inequality identically. Carroll *et al.* (1984) use a penalty function approach to fitting such a model.

Example 7.3

Carroll *et al.* (1984) illustrate their algorithm on some data collected by De Sarbo and Rao (1983) involving ten brands of pain relievers sold over the counter for remedying three common illnesses, namely headache, fever and muscle aches. The data are thus three 10×10 proximity matrices. The resulting ultrametric tree representation, accounting for 60% of the variance in the data, is shown in Fig. 7.9. The three trees are constrained to have the same topology, but the weights assigned to their nodes are allowed to vary from one illness to another. The trees reveal three distinct clusters. One cluster contains *Datril* and *Tylenol*, two asprin substitutes containing acetaminophen. Another cluster contains *Exedrin, Anacin, Bayer* and *Bufferin*, the market share leaders for aspirins. The remaining cluster consists of the less popular brands *Hudson, Cope, Ascriptin* and *Vanquish*. The strength with which the objects in these clusters group together varies from one illness to another. □

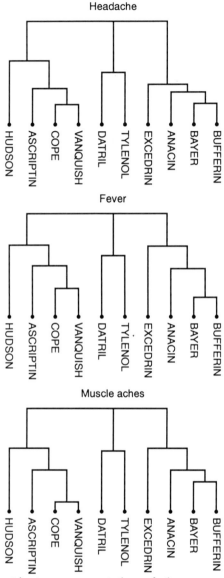

Figure 7.9 Ultrametric tree representation of three-way, two-mode data on remedies for common illnesses. (Reproduced with permission from De Soete and Carroll, 1996.)

Two-way two-mode data

7.7.2   De Soete and Carroll (1996) credit Furnas (1980) with first suggesting how the relationship between the row and column elements of a two-way two-mode proximity matrix might be portrayed using either an ultrametric or additive tree with $m + n$ terminal nodes, where m and n are, respectively, the number of rows and columns of the matrix. The weights attached to the nodes (of an ultrametric tree) or to the links (of an additive tree) are chosen so that the between-set tree distances closely correspond to the elements of the observed rectangular dissimilarity matrix. The approach is often called *tree unfolding* because of its analogy to Coombs' (1964) unfolding approach to analysing preference data via joint spatial representations (see Chapter 6). Algorithms for fitting such a model are described in De Soete *et al.* (1984), Eckes and Orlik (1991) and Espejo and Gaul (1986). Since non-symmetric two-way one-mode proximity data can be considered as a special case of two-way two-mode data where the two modes happen to correspond to the same set of stimuli, the tree unfolding approach can be used to produce ultrametric or additive trees to represent such square asymmetric data. In such a representation, each stimulus will appear twice, once as a row element and once as a column element.

Example 7.4

To illustrate the fitting of a tree model to asymmetric proximity data, citation data for 12 psychological journals given by Weeks and Bentler (1982) will be used. The data are shown in Table 7.2. The ultrametric tree representation of the

Table 7.2 Journal citation data. Citation data: Rows represent journals giving citation; columns represent citations received

	1	2	3	4	5	6	7	8	9	10	11	12
1	31	10	10	1	36	4	1	119	2	14	39	0
2	7	235	55	0	13	4	65	25	3	50	31	0
3	16	54	969	28	15	21	89	62	16	149	141	16
4	3	2	30	310	0	8	5	7	6	71	14	0
5	4	0	2	0	386	0	2	13	1	22	35	1
6	1	7	61	10	2	100	6	5	4	18	9	2
7	0	105	55	7	3	10	331	3	19	89	22	8
8	9	20	16	0	32	6	1	120	2	18	46	0
9	2	0	0	0	0	6	0	6	152	31	7	10
10	23	46	124	117	138	7	86	84	62	186	90	7
11	9	2	21	6	3	0	0	51	30	32	104	2
12	0	7	14	4	0	0	24	3	95	46	2	56

Journal
 1. *American Journal of Psychology*
 2. *Journal of Abnormal Psychology*
 3. *Journal of Personality and Social Psychology*
 4. *Journal of Applied Psychology*
 5. *Journal of Comparative and Physiological Psychology*
 6. *Journal of Consulting and Clinical Psychology*
 7. *Journal of Educational Psychology*
 8. *Journal of Experimental Psychology*
 9. *Psychometrika*
10. *Psychological Bulletin*
11. *Psychological Review*
12. *Multivariate Behavioral Research*

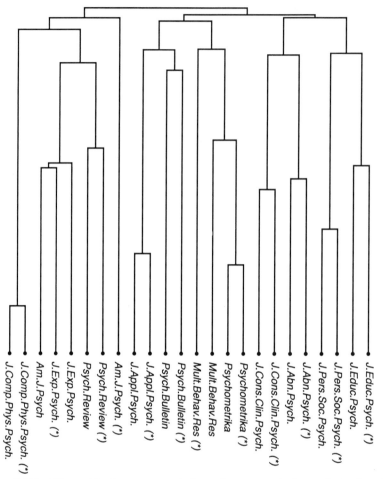

Figure 7.10 Ultrametric tree representation of asymmetric journal citation data, cited journals are indicated with an asterisk. (Reproduced with permission from De Soete and Carroll, 1996.)

data is shown in Fig. 7.10. Three major clusters emerge:

(1) *Am. J. Psych., J. Comp. Phys. Psych., J. Exp. Psych., Psych. Review,*
(2) *J. Appl. Psych. Mult. Behav. Res., Psych. Bull., Psychometrika,*
(3) *J. Abn. Psych., J. Cons. Clin. Psych., J. Educ. Psych., J. Pers. Soc. Psych.*

The first cluster consists of the 'hard' area journals in psychology, while the third cluster groups the 'soft' area journals. The second cluster comprises journals that publish mainly methodological articles. The height at which two nodes representing the same journal (once as citing journal and once as cited journal) join, displays the relative amount of self-citation, which clearly varies from journal to journal. Some interesting non-symmetric relations are apparent. *Multivariate Behavioural Research* and *Psychometrika*, for example, publish similar types of paper, but the former cites the latter more often than itself or any other journal. *Psychometrika*, on the other hand, primarily cites itself. This asymmetry can be attributed to *Psychometrika's* more prestigious standing, and its tendency to publish basic methodological papers, which are then cited by the more applied papers generally published in *Multivariate Behavioural Research.*   □

Spatial versus tree representations of proximity data

7.8   A number of authors have compared spatial and tree representations of the same data. Sattath and Tversky (1977), for example, compare the fit of additive tree representations with scaling solutions for several data sets, including the animal data of Henley (1969). In general they find that the former provided a better account of the data than the latter, as measured by a number of fit criteria. Furthermore, while the clusters produced by the trees were readily interpretable, the dimensions that emerged from the spatial representation were often more problematic. The two-dimensonal scaling solution for Henley's animal data, for example, shown in Fig. 7.11 has a stress value of 0.17 compared with a stress value for the additive tree representation of 0.07. The horizontal dimension of Fig. 7.11 is readily interpreted as size, but the interpretation of the vertical dimension as *ferocity* (Henley, 1969), is not so convincing.

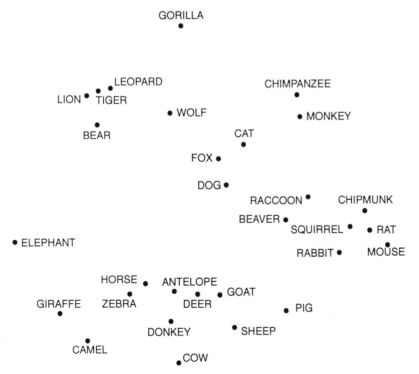

Figure 7.11   Two-dimensional scaling solution for animal data.

Pruzansky *et al.* (1982) report the results of a detailed investigation comparing spatial and tree representations of proximity data. Part of the study involved simulated data and part real data. In the simulation study, artificial data generated either by a plane or by a tree were analysed by a scaling technique and a tree fitting technique. As expected, the appropriate model fitted the data better than the inappropriate model for all levels of 'noise' introduced. Additionally, the two models were roughly comparable for all noise levels, the scaling procedure accounting for the plane data about as well as the tree model accounted for the tree data. When the inappropriate model was applied, scaling appeared to accommodate tree data slightly better than the tree model handled plane data. Two properties of the data that might help distinguish between planes and trees were explored. The first, skewness, was based on the third central moment of the distribution of distances. The second, elongation, measured the proportion of elongated triangles among all triples in a set of interpoint distance. The results of the simulation showed that trees were quite negatively skewed and became more symmetric as noise level increased. In contrast, plane data exhibited slight positive skewness. The elongation measure was higher for trees than for planes, but the difference decreased as noise level increased.

In the second part of the study by Pruzansky *et al.* (1982), both scaling and tree models were applied to 20 sets of real data. The results showed that most data sets could be clearly classified as favouring either a tree or a

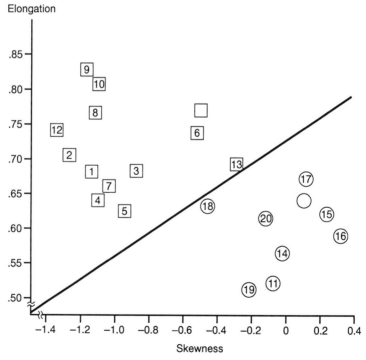

Figure 7.12   Skewness and elongation diagram. (□ = trees, ○ = planes.)

two-dimensional spatial representation. As in the simulation study, the data that were better fitted by a tree model exhibited substantial negative skewness and high elongation while the data that were better described by the scaling solution yielded very small negative or positive skewness and lower elongation. Figure 7.12, taken from Pruzansky *et al.* (1982), shows that the data sets could be separated by a straight line in the skewness and elongation plane.

These results indicate that some proximity data are better described by a tree model than by a spatial configuration, but for other data, dimensional models are more suitable. As suggested by Sattath and Tversky (1977), the appropriateness of tree or spatial representations depends largely on the structure of the stimuli involved. Emotions, for example, may be described in terms of intensity and pleasure; sound may be characterized in terms of intensity and frequency, and as such both types of stimuli are natural candidates for dimensional representations. Other sets of stimuli have a hierarchical structure that may result, for example, from an evolutionary process in which all the stimuli have an initial common structure and later develop additional distinctive features. Structures generated by an evolutionary process are likely candidates for tree representations.

Some 'hybrid' models that combine the fitting of both tree models and spatial models have been suggested by Carroll (1976), Carroll and Pruzansky (1980) and De Sarbo *et al.* (1990).

Summary

7.9 Tree models may offer an attractive alternative to the spatial models produced by MDS techniques for many sets of proximity data. Both additive and ultrametric trees can be used to produce a series of partitions of the stimuli which can often capture aspects of the data not obvious from a low-dimensional geometric configuration. Unfortunately, routine application of tree models is made difficult because of the lack of easily accessible software. Some hierarchical clustering algorithms (see Everitt, 1993) may provide adequate approximations in many cases (see Corter, 1996).

APPENDIX A

Distances in Classical Multivariate Analysis

Introduction

A.1 In the opening section of Chapter 1 the claim was made that the concept of distance is fundamental in problems involving multivariate description, multivariate inference and multivariate classification – in other words, that distance is central to multivariate analysis techniques in general. But later, in the same chapter, it was pointed out that the classical approaches to the analysis of multivariate data largely emphasize the covariance structure amongst the variables and pay little attention to the properties of the method in terms of distances between the objects or individuals on which the variables are measured. Consequently, in this appendix the distance properties of commonly used multivariate techniques will be briefly described.

Principal components analysis and classical scaling

A.2 If $\mathbf{X}$ is the $n \times p$ data matrix with column-means zero and, optionally, unit column variances, then the rotation to the principal component space is given by the $p \times p$ orthogonal matrix $\mathbf{L}$ obtained from the spectral decomposition

$$\mathbf{X'X} = \mathbf{L\Lambda L'} \tag{A1}$$

where the eigenvalues $(\lambda_1 \geqslant \lambda_2 \geqslant \ldots, \geqslant \lambda_p \geqslant 0)$ are the elements of the diagonal matrix $\mathbf{\Lambda}$.

The coordinates of the individuals in the principal components space are given as the rows of the matrix $\mathbf{Z}$

$$\mathbf{Z} = \mathbf{XL} \tag{A2}$$

The amount of variation contained in the ith principal direction is given by λ_i and since $\mathbf{L}$ is an orthogonal matrix, the Euclidean distances in the original p

dimensions are preserved in the full principal-component space, i.e.

$$d_{rs}^2 = \sum_{i=1}^{p} (x_{ri} - x_{si})^2 = \sum_{i=1}^{p} (\mathbf{x}_r' \mathbf{l}_{(i)} - \mathbf{x}_s' \mathbf{l}_{(i)})^2 \tag{A3}$$

Now let $\mathbf{L} = (\mathbf{L}_1, \mathbf{L}_2)$ where $\mathbf{L}_1$ is $p \times k$. Then $\mathbf{XL}_1$ represents a projection of the configuration $\mathbf{X}$ onto a k-dimensional sub-space spanned by the columns of $\mathbf{L}_1$. We can think of $\hat{\mathbf{X}} = \mathbf{XL}_1$ as a 'fitted' configuration in k dimensions. If we denote the distances between the rows of $\mathbf{XL}_1$ by $\hat{\mathbf{D}}$ then

$$\hat{d}_{rs}^2 = \sum_{i=1}^{k} (\mathbf{x}_r' \mathbf{l}_{(i)} - \mathbf{x}_s' \mathbf{l}_{(i)})^2 \tag{A4}$$

Consequently $\hat{d}_{rs}^2 \leqslant d_{rs}^2$ and projecting a configuration reduces the interpoint distances. The k-dimensional principal components solution is 'best' in the sense that it is the projection that minimizes the following measure of fit of $\hat{\mathbf{X}}$ for $\mathbf{X}$

$$\psi = \sum_{i=1}^{n} \sum_{j=1}^{n} (d_{ij}^2 - \hat{d}_{ij}^2) \tag{A5}$$

The similarity between principal components analysis and classical multidimensional scaling was remarked on in Chapter 3. In fact when classical scaling is applied to a matrix of Euclidean distances the resemblance becomes a formal duality since the principal component scores calculated from the covariance matrix of the data coincide with the coordinates of the scaling solution (see Gower, 1966b).

Canonical variate analysis

A.3   Suppose that the rows of the data matrix $\mathbf{X}$ have been partitioned *a priori* into g groups (for example, when the data comprise samples from g populations). Represent the number of individuals in group i by n_i, with $\sum_{i=1}^{n} n_i = n$, and let $\mathbf{x}_{ij}$ denote the vector of variable values for the jth individual in the ith group, $\bar{\mathbf{x}}_i$ denote the mean vector of the ith group and $\bar{\mathbf{x}}$ the overall mean vector. The within-group and between-group covariance matrices are defined as follows

$$\mathbf{W} = \frac{1}{n-g} \sum_{i=1}^{g} \sum_{j=1}^{n_i} (\mathbf{x}_{ij} - \bar{\mathbf{x}}_i)(\mathbf{x}_{ij} - \bar{\mathbf{x}}_i)' \tag{A6}$$

and

$$\mathbf{B} = \frac{1}{g-1} \sum_{i=1}^{g} n_i (\bar{\mathbf{x}}_i - \bar{\mathbf{x}})(\bar{\mathbf{x}}_i - \bar{\mathbf{x}})' \tag{A7}$$

respectively.

Canonical variates are linear combinations $\mathbf{z}_i = \mathbf{Xa}_i$, where the vector $\mathbf{a}_i$ is chosen to maximize the ratio of the between-group to within-group variance of the z_i, i.e.

$$\frac{\mathbf{a}_i'\mathbf{B}\mathbf{a}_i}{\mathbf{a}_i'\mathbf{W}\mathbf{a}_i} \tag{A8}$$

This maximization leads to a solution of the generalized eigenvalue/eigenvector equation

$$(\mathbf{B} - l_i\mathbf{W})\mathbf{a}_i = \mathbf{0} \tag{A9}$$

for which there are, in general, $s = \min(g-1, p)$ non-zero eigenvalues, $l_1 > l_2 > \dots l_2$ and $\mathbf{a}_i$ is the eigenvector corresponding to l_i. If we write $\mathbf{A} = (\mathbf{a}_1, \dots, \mathbf{a}_s)$ and $\mathbf{L} = \text{diag}(l_1, \dots, l_s)$, then (A9) implies that $\mathbf{BA} = \mathbf{WAL}$. The coefficients $\mathbf{a}_i$ are generally normalized so that $\mathbf{A}'\mathbf{WA} = \mathbf{I}$. The results of a canonical variate analysis are generally displayed by plotting the canonical variate group means $\bar{\mathbf{z}}_i = \mathbf{A}'(\bar{\mathbf{x}}_i - \bar{\mathbf{x}})$ against orthogonal axes. When all s canonical variate axes are used, the squared Euclidean distance between points representing two means in the canonical variate space is equal to the squared Mahalanobis distance between the corresponding means in the original data space. Moreover, the first r dimensions of the canonical variate space yield the best r-dimensional representation of the groups in the sense of maximizing the between-group relative to the within-group variability of the data.

Krzanowski (1994) points out that although the above account of canonical variate analysis is the one usually provided in standard text books, an earlier approach (Rao, 1948) seeks the r-dimensional space in which the total Mahalanobis square distance between pairs of samples is maximized. Ashton *et al.* (1957) argued that this objective is the more appropriate one for descriptive analysis of data and pointed out that it can be achieved by using the unweighted between-group covariance matrix

$$\mathbf{B}_u = \frac{1}{g-1} \sum_{i=1}^{g} (\bar{\mathbf{x}}_i - \bar{\mathbf{x}})(\bar{\mathbf{x}}_i - \bar{\mathbf{x}})' \tag{A10}$$

in place of the weighted version, $\mathbf{B}$ in (A7); the more disparate the sample sizes n_i the more the two solutions will differ. With this approach Gower (1966a) shows that there is again a formal duality between canonical variate analysis and classical scaling in the sense that the canonical variate group means are identical to the coordinates found from classical scaling applied to the matrix of squared Mahalanobis distances between each pair of groups.

Example A1

To illustrate this procedure, the data shown in Table A1, taken from Nathanson (1971), will be used. The table shows the squared Mahalanobis distances between ten types of galaxies calculated from seven descriptive parameters such as diameter, surface brightness, colour, etc., observed on 273 galaxies. The canonical variate means in two dimensions can be found by simply applying classical scaling to this 10×10 distance matrix. The resulting coordinates are plotted in Fig. A1. □

Table A1   Inter-group Mahalanobis distances for ten types of galaxies

	1	2	3	4	5	6	7	8	9	10
1	0.00									
2	3.29	0.00								
3	2.79	1.13	0.00							
4	3.52	1.75	1.45	0.00						
5	3.77	2.97	1.71	2.02	0.00					
6	3.27	3.01	2.13	1.89	1.27	0.00				
7	3.93	3.72	3.00	2.25	1.86	0.68	0.00			
8	3.86	5.12	4.11	3.24	3.15	1.59	1.51	0.00		
9	3.77	5.70	4.85	3.85	3.41	1.74	2.05	0.91	0.00	
10	4.12	6.88	6.02	7.03	5.38	4.09	4.03	2.24	1.87	0.00

1. Irregulars(I); 2. SBc, 3. Sc; 4. Sbb; 5. Sb; 6. Sba; 7. Sa; 8. SBO; 9. SO; 10. Ellipticals (E).

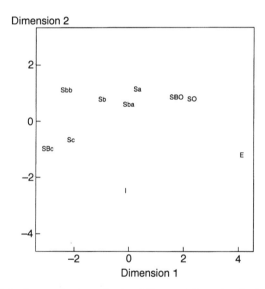

Figure A1.   Plot of coordinates obtained by applying classical multidimensional scaling to the distance matrix shown in Table A1.

Correspondence analysis

A.4   Correspondence analysis is essentially a technique for displaying contingency table data in the form of a scatterplot or 'map' in which row and column categories are represented as points. If the frequencies in the contingency table are denoted by $n_{ij}, i = 1, \ldots r, j = 1, \ldots c$, the Pearsonian residual for cell ij is defined as

$$e_{ij} = \frac{(n_{ij} - n_{i.}n_{.j}/n)}{n_{i.}n_{.j}/n} \tag{A11}$$

where $n_{i.}$ and $n_{.j}$ represent row and column marginal totals and n is the grand total. Correspondence analysis is usually formulated as involving the singular value description of the matrix **E** with elements e_{ij}; for example, the following is taken from Krzanowski and Marriott (1994)

$$\mathbf{E} = \mathbf{UDV}'$$ (A12)

where **D** is the diagonal matrix containing the square roots of the non-zero eigenvalues of either **E'E** or **EE'** (ranked in descending order) and **U**, **V** are orthogonal matrices whose columns are the corresponding eigenvectors (normalized to have unit sums of squared elements) of **EE'** and **E'E** respectively. The coordinates of the row profiles of the table are given by the rows $\mathbf{f}_i$ of $\mathbf{F} = \mathbf{D}_r^{-1/2}\mathbf{UD}$ while the coordinates of the column profiles are given by the rows $\mathbf{g}_i$ of $\mathbf{G} = \mathbf{D}_c^{-1/2}\mathbf{VD}$. The pairs of row and column coordinates $\mathbf{f}_i$, $\mathbf{g}_i$ are the elements in the orthogonal decomposition of the residuals from the independence model in decreasing order of importance.

Correspondence analysis can also be described in terms of the classical scaling of two particular distance matrices, one for the rows of the contingency table and one for the columns. For the columns, for example, the distances are derived from the table with elements $p_{ij} = n_{ij}/n_{.j}$ as

$$d_{ij}^{(\text{columns})} = \left(\sum_{k=1}^{k} \left[\frac{(p_{ki} - p_{kj})^2}{p_{k.}} \right] \right)^{1/2}$$ (A13)

where $p_{k.} = n_{k.}/n$; $k = 1, \ldots, r$. This chi-squared distance, by dividing each squared difference by the expected frequency, effectively compensates for the different levels of occurrence of the categories. More formally, the chi-squared distance can be justified as a way of standardizing variables under a multinomial or Poisson distributional assumption, and is, in many respects, the natural choice for frequency data. For a contingency table with r rows and c columns, it can be shown that the chi-squared distances can be represented exactly in $\min\{r - 1, c - 1\}$ dimensions.

Example A2

As an example consider the data shown in Table A2 concerned with the influence of age on relationships with boyfriends. Each of 139 girls has been classified into one of three groups – no boyfriend, boyfriend/no sexual intercourse, boyfriend/sexual intercourse. In addition, the age of each girl has been recorded leading to the frequencies in the cross-tabulation of type of relationship against age shown in Table A2. The chi-squared distances between the age

Table A2 The influence of age on relationships with boyfriends

	Age				
	15	16	17	18	19
No boyfriend	21	21	14	13	8
Boyfriend/no sexual intercourse	8	9	6	8	2
Boyfriend/sexual intercourse	2	3	4	10	10

groups found as described above are as follows

$$\mathbf{D}^{(\text{age groups})} = \begin{array}{c} \\ 1 \\ 2 \\ 3 \\ 4 \\ 5 \end{array} \begin{array}{ccccc} 1 & 2 & 3 & 4 & 5 \\ \left(\begin{array}{ccccc} 0.001 & & & & \\ 0.09 & 0.00 & & & \\ 0.26 & 0.19 & 0.00 & & \\ 0.66 & 0.59 & 0.41 & 0.00 & \\ 1.07 & 1.01 & 0.83 & 0.51 & 0.00 \end{array} \right) \end{array} \tag{A14}$$

Applying classical multidimensional scaling to this matrix results in the following two-dimensional coordinates for the five age groups.

Age group	c1	c2
1	−0.40	0.06
2	−0.34	0.00
3	−0.15	0.00
4	0.23	−0.15
5	0.67	0.09

Since $\min\{r-1, c-1\}$ for this table is 2, the Euclidean distances between points with these coordinates values reproduce exactly the chi-squared distances given above.

A similar analysis could be applied to derive coordinate values for the rows of the table. □

An attempt at fully integrating multidimensional scaling and multivariate analysis has been made by Meulman (1986, 1992). This formulation assumes that the variables in a set of multivariate data are functions defined on the individuals and may be non-linearly transformed. From the optimally transformed variables, distances between individuals are approximated through a metric MDS approach. An optimal transformation is that particular choice minimizing the fit function used to measure the discrepancy between the observation space and the representation space. If a perfect relationship exists between the given observation space and a representation space, there is no need for a transformation. Meulman defines a general least squares fit function that encompasses many multivariate analysis methods. From a geometrical viewpoint, variable transformation involves the replacement of the coordinates of the individuals in the observation space by some other coordinate system, under the restriction of a possibly non-linear relation between the coordinates in the two systems. The process of including transformation of the variables can be motivated in a number of different ways. It may be known that the variables have not been measured on an interval scale, and therefore an analysis should be invariant under functions that are more general than linear transformations, such as monotonic or non-monotonic transformations of the data. For example, in a nominal (non-monotonic) transformation, only categorical information from the original variable is retained; if ordinal information is to be maintained, the original variable is transformed monotonically. In addition, non-linear transformations of variables may reveal non-linear relationships between the original variables.

Meulman (1986) makes the important point that it is not always better to use distance methods for multivariate data. The distance approach is only one of a number of possible points of view and when the prevailing interest is not in the objects it would clearly not be sensible to choose an analysis that is based on distance functions derived for the objects. When the primary interest is in the study of relationships between the variables, conventional multivariate analysis, with possible variable transformation, would be more appropriate.

APPENDIX B

Software for Multidimensional Scaling

Introduction

B.1 So far we have described multidimensional scaling in terms of models, their applications and interpretation. We have said a little about the strategies or algorithms used to fit these models to data, but we have completely avoided mentioning any particular programs. This was mainly to avoid littering the chapters with acronyms! One purpose of this appendix is to introduce the acronyms so that the reader can make a connection between the models described in this book and the program names used in the multidimensional scaling literature. Another purpose of this appendix is to list some of the packages and programs currently available for multidimensional scaling.

A brief history of programs for MDS

B.2 The first approach to MDS was classical metric scaling (Section 3.2). The algorithm for classical scaling was developed by Torgerson (1952, 1958) and implemented by Young (Young and Torgerson, 1967) in a program called TORSCA (TORgerson SCAling).

Shepard (1962a,b) developed the first program for non-metric scaling (Section 3.3). His ideas were extended by Kruskal (1964a,b) whose program MDSCAL (MultiDimensional SCALing) minimizes stress by iterating through monotonic least squares regression to update the disparities, followed by steepest descent to update the configuration of points. Young's program TORSCA was also improved to cope with non-metric analyses (Young and Torgerson, 1967). Subsequently, Kruskal *et al.* (1977) combined the best features of TORSCA and MDSCAL into KYST (Kruskal, Young, Shepard, Torgerson).

Another approach to non-metric MDS was due to Guttman (1968), who prefers the name 'smallest space analysis'. Guttman defined a loss function called the coefficient of alienation which is basically equivalent to stress but leads to a different algorithm. Lingoes (1973) implemented this method in a series of programs called SSA-I through SSA-IV.

Lingoes and Roskam (1973) then created MINISSA (Michigan–Israel–Nijmegen Integrated Smallest Space Analysis) which merges the two main traditions of basic non-metric MDS: the Shepard–Kruskal approach and the Guttman–Lingoes approach. The program can be used for both Euclidean and Minkowski distances. Roskam subsequently wrote MRSCAL (MetRic SCA-Ling), which is the metric counterpart of MINISSA and allows both linear and logarithmic transformation of the data.

The following programs were developed for rectangular data (Chapter 6). Lingoes wrote a series of programs called SSAR-I through SSAR-IV for non-metric scaling of rectangular data. MINIRSA (MINI Rectangular Smallest Space Analysis) was created by Roskam in 1973 for row-conditional non-metric unfolding. MDPREF (MultiDimensional PREFerence scaling) by Carroll and Chang is the vector-model counterpart to MINIRSA, but for interval data.

Carroll and Chang developed PROFIT (PROperty FITting) for external analysis (Section 6.3.5), of the row conditional vector model, as well as PREFMAP (PREFerence MAPping) which is capable of fitting the entire hierarchy of models described in Carroll (1972) (see Section 6.3.4). Both metric and non-metric analyses are possible in PROFIT and PREFMAP. POLYCON (POLYnomial CONjoint analysis) is a very general program developed by Young (1972) which can analyse both square and rectangular matrices metrically and non-metrically and allows a general functional relationship between coordinates and disparities, for example inner products, Euclidean distances or any other form of distance function.

The first program for fitting the weighted Euclidean model (Section 5.2) was INDSCAL (INdividual Differences SCALing), created by Carroll and Chang (1970), who invented this method of carrying out individual differences scaling. INDSCAL implements the inner products approach to fitting the model (Section 5.2.1). SINDSCAL is a streamlined version developed by Pruzansky (1975) and IDIOSCAL (Individual Differences in Orientation SCALing) was written by Carroll and Chang (1972) to implement the rotated and weighted Euclidean model (Section 5.3). Harshman (1972) independently implemented a similar procedure to Carroll and Chang in a program called PARAFAC (PARAllel FACtors). Tucker (1972) wrote a similar program for 'three-mode three-way' analysis. Other programs for fitting the weighted Euclidean model include Ramsay's program MULTISCALE (Ramsay, 1978b, 1991) for maximum likelihood estimation (Section 5.5) and PINDIS (Procrustean INdividual DIfferences Scaling) by Lingoes and Borg (1977) (Section 5.4).

ALSCAL (Alternating Least squares SCALing) was created by Takane, *et al.* (1977). The program fits the weighted Euclidean and the rotated and weighted Euclidean model by minimizing S-Stress as described in Section 5.2.3. The algorithm is very general and can be used for metric or non-metric scaling, for row or matrix conditional data, for asymmetric and rectangular data, as well as replicated versions of all these models and three-way unfolding. Missing values are allowed and internal as well as external analyses are possible.

One problem with the steepest descent algorithm which most of the programs mentioned so far are based on is that the gradient of the stress function is not defined when two points coincide, which happens regularly when $q = 1$. De

Leeuw and Heiser have therefore developed a program called SMACOF which uses a majorization algorithm (de Leeuw and Heiser, 1977; de Leeuw, 1988). In order to avoid getting trapped in local minima, Groenen and Heiser (1996) combined the majorization algorithm with a tunnelling method. When a local minimum has been reached, the program searches for another configuration with the same value of stress and looks for the local minimum in that region until the lowest local minimum has been found.

Most of the programs described in this section are tabulated in Table B.1.

Statistical packages and libraries incorporating MDS

B.3 An increasing number of statistical packages include options for multi-dimensional scaling. In this section we only mention the larger packages and libraries. Table B.2 gives a list of programs from Table B.1 that are currently available in a similar form, together with the package or library in which the program is available or, in the case of stand-alone programs, the name of the person or organization from which the program may be obtained.

SPSS

B.3.1 SPSS is a general-purpose statistical package, oriented towards the needs of social scientists and maintained by SPSS Inc. (see http://www.spss.com). This package uses the program ALSCAL for multidimensional scaling (SPSS, 1994).

A brief description of the syntax necessary to fit some of the more complex models might be helpful; so, for example, the first-way weights model (Section 6.2.2) may be obtained by specifying /shape=asymmetric and /model= ASCAL. The rotated and weighted Euclidean model (Section 5.3) may be obtained by specifying /model=gemscal and /criteria= directions(n) where *n* is the number of principal directions. For an unfolding analysis, simply set /shape=rec. External unfolding analyses are achieved by also specifying /file=[file containing configuration] Rowconf(fix).

One mistake that tends to be made when first running ALSCAL is to list the variables in an order that is different from the order of the corresponding columns in the data matrix (for example, in alphabetical order) which leads to nonsensical results.

SAS

B.3.2 SAS is a general-purpose software package that has procedures for a wide variety of statistical analyses (see http://www.sas.com). The main procedure for multidimensional scaling is Proc MDS. This procedure combines most of the features of two programs, proc MLSCALE by Ramsay (the SAS version of MULTISCALE) and proc ALSCAL by Young (the SAS version of ALSCAL) from the SUGI Supplemental library (SUGI, 1986), a set of user-supplied functions not specifically supported by SAS. By specifying

Table B.1 Computer programs according to the number of ways and modes, symmetry, underlying model and corresponding section or figure. Each program is followed by up to three symbols in brackets: [m (metric), n (non-metric) or b (both)] for the level of measurement, followed by [I (internal), E (external) or B (both)] and if the program allows distances other than Euclidean, a third symbol appears [M (Minkowski) or C (city block)]. The sections and figures are underlined if they correspond exactly to the program name on the right

Ways	Modes	Sym	Model	Section (Figure)	Program
2	1	sym	classical scaling	3.2 (3.2, 3.3) (3.12, 3.15, 3.16)	TORSCA(m,I)
			metric/non-metric scaling	3.3, 3.4, 3.5 (<u>3.11</u>) (3.4, 3.7, 3.8, 3.9) (3.10)	ALSCAL(b,B) KYST(b,B,M) MINISSA(n,I,M) MDSCAL(n,I,M) MRSCAL(m,I,M) MULTISCALE(m,I) POLYCON(b,B,M) SSA-I(n,I,C) TORSCA(m,I)
			vector model (factor analysis)		SSA-III(n,I,C) POLYCON(b,B,M)
		asym	first way weights	6.2.2 (<u>6.3</u>)	ALSCAL(b,B)
			vector model	6.2.3	DEDICOM(m,I)
	2	–	unfolding (distance model)	6.3.1 (<u>6.5 6.7</u>) (6.4) 6.3.3 <u>6.3.5</u> (<u>6.9</u>)	ALSCAL(b,B) KYST(b,B,M) MINIRSA(n,I) PREFMAP(b,E) POLYCON(b,B,M) SSAR-I(n,I,C) SSAR-II(n,I,C)
			weighted unfolding general unfolding	6.3.4 6.3.4	PREFMAP(b,E) PREFMAP(b,E)
			vector model	6.3.2 (<u>6.8</u>) 6.3.3 <u>6.3.5</u>	MDPREF(m,I) POLYCON(b,B,M) PREFMAP(b,E) PROFIT(b,E)
3	2	sym	weighted Euclidean model	5.2, <u>5.2.1</u>, (<u>5.1</u>, <u>5.2</u>) <u>5.2.3</u> (<u>5.3</u>, <u>5.4</u>) (<u>5.5</u>, <u>5.8</u>) <u>5.5</u> (<u>5.12</u>) <u>5.4</u>	(S)INDSCAL(m,B) ALSCAL(b,B) MULTISCALE(m,I) PINDIS(b,B)
			rotated and weighted Euclidean model	5.3 (<u>5.9</u>, <u>5.10</u>) <u>5.4</u>	ALSCAL(b,B) IDIOSCAL(m,B) PINDIS(b,B)
			vector weighted model	<u>5.4</u> (<u>5.11</u>)	PINDIS(b,B)
		asym	general Euclidean model	6.4	ALSCAL(b,B)
	3	–	general Euclidean model	6.4	ALSCAL(b,B)

Table B.2 Program acronyms, references in which the algorithm is described and the package or source where the program or a similar version of it can be obtained. Note that some packages do not use any program in its original form; this is indicated by using a general heading under one of the appropriate programs, for example 'Kruskal–Shepard non-metric' under 'KYST'

Program	References	Package/source
ALSCAL (individual differences alternating least squares)	(Takane et al., 1977)	F.W. Young[1] SPSS SAS IMSL
TORSCA (classical scaling)	(Torgerson, 1958)	GENSTAT S-PLUS NAG SOLO T.F. Cox and M.A.A. Cox[1]
KYST (Kruskal-Shepard non-metric scaling)	(Kruskal, 1964a)	Bell-Labs[1] SYSTAT SOLO isoMDS[1] STATISTICA NAG MDS(X) T.F. Cox and M.A.A. Cox[1]
INDSCAL	(Carroll and Chang, 1970)	Bell-Labs[1] MDS(X) T.F. Cox and M.A.A. Cox[1]
IDIOSCAL	(Carroll and Chang, 1972)	Bell-Labs[1]
MDPREF	(Carroll and Chang, 1972)	Bell-Labs[1] T.F. Cox and M.A.A. Cox[1] MDS(X)
MDSCAL	(Kruskal and Wish, 1978)	see KYST
MINISSA	(Roskam and Lingoes, 1970)	MDS(X) T.F. Cox and M.A.A. Cox[1] see KYST and SSA SYSTAT 7
MINISRA	(Davies and Coxon, 1983)	MDS(X) T.F. Cox and M.A.A. Cox[1]
MULTISCALE (max. likelihood 3 way scaling)	(Ramsay, 1978a)	J.O. Ramsay[1] SAS
MRSCAL	(Davies and Coxon, 1983)	MDS(X) see KYST and SSA
PREFMAP	(Carroll and Chang, 1972)	MDS(X) SYSTAT 7
PINDIS	(Lingoes and Borg, 1978)	MDS(X)
POLYCON	(Young, 1972)	F.W. Young[1]
PROFIT	(Carroll and Chang, 1973)	Bell-Labs MDS(X)
SSA/SSAR (Guttman–Lingoes) non-metric scaling)	(Guttman, 1968)	FACET[1] SYSTAT STATISTICA

[1] see Section B.4.

fit=squared and formula=1 (except for unfolding where formula 2 is used), Proc MDS can be made to produce results very similar to ALSCAL. All the various models generally available in ALSCAL are available in Proc MDS. Running the procedure with the options proc mds fit=log level= loginterval . . .; produces results very similar to MLSCALE. It is however not possible to allow the variances to differ between subjects. Proc MDS is described in Chapter 17 of SAS (1996).

SYSTAT

B.3.3 SYSTAT is a general-purpose statistics package maintained by SPSS Inc. (See Section B.3.1). The function MDS offers the Guttman and Kruskal methods of scaling for general Minkowsi distances. The Kruskal method minimizes stress (formula 1) or S-Stress and the Guttman method minimizes the coefficient of alienation. Unfolding and individual differences models are available for all three options of loss functions. The relationship between distances and dissimilarities may be specified as monotonic, linear, log and/or power (Steven's power function). The function MDS is described in SYSTAT (1996). SYSTAT also allows the fitting of additive trees and version 7 has a function for 'perceptual mapping' which includes MDPREF, PREFMAP and Procrustes rotation.

SOLO (BMDP)

B.3.4 SOLO Categorical Data Analysis is a software package written by J. L. Hintze (SOLO, 1991) and distributed by BMDP. Classical scaling and Kruskal's method of non-metric scaling are available.

GENSTAT

B.3.5 Genstat is a general statistics package developed by statisticians at the Rothamsted Experimental Station in the UK. It is now maintained by NAG (see Section B.3.8). The MDS directive is based on the work of L. G. Underhill. The method options are either non-metric (i.e. using monotonic regression) or linear (i.e. using linear regression through the origin) to compute the fitted distances with three possible measures of stress: stress, squared-stress and log-stress. There is also a ties option: primary, secondary or tertiary. In addition, Genstat has the principal coordinates directive which provides classical scaling. The directives are described in Genstat 5 (1993).

STATISTICA

B.3.6 STATISTICA is a general purpose statistics package by StatSoft Inc. (see http://www.statsoftinc.com). The multidimensional scaling module includes a full implementation of (non-metric) multidimensional scaling. The program employs an iterative procedure to minimize the stress value or the coefficient of alienation.

IMSL

B.3.7 The IMSL STAT/LIBRARY is a Fortran subprogram library for solving problems in statistical analysis (see http://www.vni.com). It contains a number of subroutines for multidimensional scaling which are described in Chapter 14 of IMSL (1987).

The subroutines relevant to multidimensional scaling are MSIDV, an individual differences module, similar to ALSCAL but for metric data, MSDST which computes distance matrices based upon a model, MSSTN to standardize the input data, MSDBL to double centre a dissimilarity matrix, MSINI to compute initial estimates and MSTRS to compute stress given disparities and distances.

NAG

B.3.8 The Numerical Algorithms Group (NAG) is a Not-For-Profit software house (see http://www.nag.co.uk). Among its products are numerical subroutine libraries in various computer languages. The subroutine G03FCF of the FORTRAN 77 library, developed by Adrian Cook, performs multidimensional scaling using monotonic regression and either stress or squared stress. The coordinates, stress and distances are returned. The subroutine G03FAF performs classical scaling. Both subroutines are described in NAG (1993).

Individual programs for MDS

B.4 There are many stand-alone programs that have not been incorporated into many (if any) major packages or libraries in similar form. Examples include PINDIS and PREFMAP. Many of these programs are available from the Internet. In this section, we list the sources that we are aware of at the moment (many of them were referred to in Table B.1). Addresses frequently change, especially on the Internet, and we cannot guarantee that this information will be accurate in the future.

MDS(X): PINDIS, PREFMAP, INDSCAL etc.

B.4.1 The MDS(X) Series of multidimensional scaling programs and documentation (Davies and Coxon, 1983) can be obtained from

Program Library Unit
University of Edinburgh
18 Buccleuch Place
Edinburgh EH8 9LN, UK

The programs are available as FORTRAN source files on a magnetic disk.

J. O. Ramsay: MULTISCALE

B.4.2   The program MULTISCALE can be obtained, together with the manual from

J. O. Ramsay
Department of Psychology
McGill University
Stewart Biological Sciences Building
1205 Dr. Penfield Avenue
Montreal, QC, Canada H34 1B1
jramsay@psych.mcgill.ca

The program may also be downloaded by anonymous ftp from ego.psych. mcgill.ca, directory pub/ramsay/multiscale.

F. W. Young: ALSCAL and VISTA

B.4.3   The FORTRAN and compiled PC versions of ALSCAL can also be obtained from http://forrest.psych.unc.edu/research/ALSCAL.html. A manual for ALSCAL can also be obtained.

B. D. Ripley: isoMDS

B.4.4   isoMDS is an S-Plus function for Kruskal's non-metric MDS. It can be obtained from http://www.stats.ox.ac.uk/pub or from http://lib.stat. cmu.edu/S.

Bell Lab: KYST, INDSCAL, IDIOSCAL. MDPREF, PROFIT etc.

B.4.5   Many programs written at Bell Labs (for example by Carroll and Chang) can be obtained by anonymous ftp from netlib.att.com (directory /netlib/mds) or by sending an e-mail to netlib@research.att.com, initially with the message 'send readme index from mds' to see what is available.

FACET: SSA

B.4.6   FACET is a suite of programs to support facet analysis, smallest space analysis (Guttman's method of MDS), multidimensional scalogram analysis and partial order scalogram analysis. FACET is available from http://www. canterbury.ac.nz/psyc/barrett/programs.html.

G. J. Sandell: mdsplot

B.4.7   An S function for plotting three-dimensional MDS configurations. It is available from http://lib.stat.cmu.edu/snews/Burst/5382 and was used for plotting the bodypart diagrams in Chapter 5.

T. F. Cox and M.A.A. Cox: CLSCAL, INDSCAL, UNFOLDIN etc.

B.4.8   Programs on a disk supplied with the book by Cox and Cox (1994).

Many thanks to all the members of the 'allstat' mailing list who replied to our request for information on software for multidimensional scaling.

References

Anderson, T.W. and Bahadur, R.R. 1962: Classification into two multivariate normal populations with different covariance matrices. *Annals of Mathematical Statistics*, **33**, 420–31.

Arabie, P. 1991: Was Euclid an unnecessarily sophicated psychologist? *Psychometrika*, **56**, 567–87.

Ashton, E.H., Healy, M.J.R. and Lipton, S. 1957: The descriptive use of discriminant functions in physical anthropology. *Proceedings of Royal Society, Series B*, **146**, 552–72.

Attneave, F. 1950: Dimensions of similarity. *American Journal of Psychology*, **63**, 516–56.

Balakrishnan, V. and Sanghvi, L.D. 1968: Distance between populations on the basis of attribute data. *Biometrics*, **24**, 859–65.

Barlow, R.E., Bartholomew, D.J., Bremner, J.M. and Brunk, H.D. 1972: *Statistical Inference under Order Restrictions*. New York: Wiley.

Bartholomew, D.J. 1959: A test of homogeneity for ordered alternatives. *Biometrika*, **46**, 36–48.

Bennett, J.F. and Hayes, W.L. 1960: Multidimensional unfolding: determining the dimensionality of ranked preference data. *Psychometrika*, **25**, 27–43.

Berge, J.M.F., Kiers, H.A.L. and Krijnen, W.P. 1993: Computational solutions for the problem of negative saliences and nonsymmetry in INDSCAL. *Psychometrika*, **10**, 115–24.

Bloxom, B. 1968: Individual differences in multidimensional scaling. *Research Bulletin*, pp. 68–54. Princeton, NJ: Educational Testing Service.

Bookstein, F.L. 1978: *The Measurement of Biological Shape and Shape Change. Lecture Notes in Biomathematics*. Vol. 24. Berlin: Springer.

Bookstein, F.L. 1991: *Morphometric Tools for Landmark Data: Geometry and Biology*. Cambridge: Cambridge University Press.

Borg, I. and Lingoes, J.C. 1978: What weights should weight have in individual differences scaling? *Quality and Quantity*, **12**, 233–37.

Borg, I. and Lingoes, J.C. 1980: A model and algorithm for multidimensional scaling with external constraints on the distances. *Psychometrika*, **45**, 25–38.

Brandeau, M.L. and Chiu, S.S. 1988: Parametric facility location in a tree network with an L_p norm cost function. *Transportation Science*, **22**, 59–69.

Bray, J.R. and Curtis, J.T. 1957: An ordination of the upland forest communities of S. Wisconsin. *Ecological Monographs*, **27**, 325–49.

Brindley, G.S. and Lewin, W.S. 1968: The sensations produced by electrical stimulation of the visual cortex. *Journal of Physiology*, **196**, 479–93.

Brossier, G. 1985: Approximation des dissimilarities par des arbes additifs. *Mathematiques et Sciences Humaines*, **91**, 5–21.

Buneman, P. 1971: The recovery of trees from measures of dissimilarity. *Mathematics in the Archeological and Historical Sciences*, F.R. Hodson, D.G. Kendall and P. Tautu, (Eds), pp. 387–95. Edinburgh: Edinburgh University Press.

Busemann, H. 1950: The foundations of Minkowskian geometry. *Commentaru Mathematici Helvetia*, **24**, 156–86.

Busemann, H. 1955: *The Geometry of Geodesics*. New York: Academic Press.

Caillez, F. 1983: The analytical solutions to the additive constants problem. *Psychometrika*, **48**, 305–08.

Carmichael, J.W., George, J.A. and Julius, R.S. 1968: Finding natural clusters. *Systematic Zoology*, **17**, 144–50.

Carroll, J.D. 1972: Individual differences and multidimensional scaling. *Multidimensional Scaling*, Vol. 2. R.N. Shepard, A.K. Romney and S.B. Nerlove, (Eds), pp. 105–53. New York: Seminar Press.

Carroll, J.D. 1976: Spatial, non-spatial and hybrid models for scaling. *Psychometrika*, **41**, 439–63.

Carroll, J.D. and Arabie, P. 1980: Multidimensional scaling. *Annual Reviews of Psychology*, **31**, 607–49.

Carroll, J.D. and Chang, J.J. 1970: Analysis of individual differences in multidimensional scaling via an N-way generalisation of 'Eckart-Young' decomposition. *Psychometrika*, **35**, 283–319.

Carroll, J.D. and Chang, J.J. 1972: IDIOSCAL: a generalisation of INDSCAL allowing idiosyncratic reference systems as well as an analytic approximation to INDSCAL. *Paper presented at Psychometric Society*, Princeton, NJ.

Carroll, J.D. and Chang, J.J. 1973: A method for fitting a class of hierarchical tree structure models to dissimilarity data and its application to some body parts data of Miller's. *Proceedings of the 81st Annual Convention of the American Psychological Association*, **8**, 1097–98.

Carroll, J.D. and Pruzansky, S. 1975: Fitting of hierarchical tree structure (HTS) models, mixtures of HTS models and hybrid models, via mathematical programming and alternating least squares. In *U.S. Japan Seminar on Theory, Methods and Applications of Multidimensional Scaling and Related Techniques*. University of California, San Diego, La Jolla.

Carroll, J.D. and Pruzansky, S. 1980: Discrete and hybrid scaling models. *Similarity and Choice*. E.D. Lantermann and H. Feger, (Eds), pp. 108–39. Bern: Huber.

Carroll, J.D. and Wish, M. 1974: Models and methods for three-way multidimensional scaling. *Contemporary Developments in Mathematical Psychology*, Vol. 2. D.H. Krantz, R.C. Atkinson, R.D. Luce and P. Suppes, (Eds). San Francisco: W.H. Freeman.

Carroll, J.D., Clark, L.A. and De Sarbo, W.S. 1984: The representation of three-way proximity data by single and multiple tree structures models. *Journal of Classification*, **1**, 25–74.

Chaddha, R.L. and Marcus, L.F. 1968: An empirical comparison of distance statistics for populations with unequal covariance matrices. *Biometrics*, **24**, 683–94.

Chang, C.L. and Lee, R.C.J. 1973: A heuristic relaxation method for non-linear mapping in cluster analysis. *IEEE Transactions on Systems, Man and Cybernetics*, **SMC2**, 197–200.

Cliff, A.D., Haggett, P., Smallman-Raynor, M.R., Stroup, D.F. and Williamson, G.D. 1995: The application of multidimensional scaling methods to epidemiological data. *Statistical Methods in Medical Research*, **4**, 102–23.

Commandeur, J.J.F. 1991: *Matching configurations*. Ph.D. Thesis, University of Leiden.

Constantine, A.G. and Gower, J.C. 1978: Graphical representation of asymmetric matrices. *Applied Statistics*, **27**, 297–304.

Constantine, A.G. and Gower, J.C. 1982: Models for the analysis of interregional migration. *Environment and Planning A*, **14**, 477–97.

Coombs, C.H. 1950: Psychological scaling without a unit of measurement. *Psychological Review*, **57**, 148–58.

Coombs, C.H. 1964: *A Theory of Data*. New York: Wiley.

Coombs, C.H. and Smith, J.E.K. 1973: On the detection of structure in attitudes and developmental processes. *Psychological Review*, **80**, 337–51.

Corbet, G.B., Cummins, J., Hedges, S.R. and Krzanowski, W.J. 1970: The taxonomic status of British water voles. *Journal of Zoology*, **161**, 301–16.

Corter, J.E. 1996: *Tree Models of Similarity and Association*. Beverly Hills, CA: Sage Publications.

Cox, F.C. and Cox, M.A.A. 1994: *Multidimensional Scaling*. London: Chapman and Hall.

Coxon, A.P.M. 1982a: The mapping of family composition preferences: a scaling analysis. *Key Texts in Multidimensional Scaling*. P.M. Davies and A.P.M. Coxon, (Eds), London: Heinemann Educational Books.

Coxon, A.P.M. 1982b: *The User's Guide to Multidimensional Scaling*. Portsmouth: Heinemann.

Cross, D.V. 1965a: Metric properties of multidimensional stimulus generalization. *Stimulus Generalization*, pp. 72–93. Stanford: Stanford University Press.

Cross, D.V. 1965b: Multidimensional stimulus control of the discriminature response in experimental conditioning and psychophysics. Technical Report No 05613-4-F (78[d]). Ann Arbor: University of Michigan.

Davies, P.M. and Coxon, A.P.M. 1983: *MDS(X) User Manual*. Edinburgh: Inter-University Research Council Series, Report 55.

de Leeuw, J. 1988: Convergence of the majorization method for multidimensional scaling. *Journal of Classification*, **5**, 163–80.

de Leeuw, J. and Heiser, W. 1977: Convergence of correction matrix algorithms for multidimensional scaling. *Geometrical Representations of Relational Data*, J.C. Lingoes, (Ed), Ann Arbor, Michigan: Mathesis Press.

de Leeuw, J. and Heiser, W. 1982: Theory of multidimensional scaling.

Handbook of Statistics, Vol. 2: Classification, Pattern Recognition and Reduction of Dimension. P.R. Krishnaiah, and L. Kanal (Eds), pp. 285–316. Amsterdam: North-Holland.

De Sarbo, W.S. and Rao, V.R. 1983: A constrained unfolding model for produce positioning and market segmentation. Unpublished manuscript, Bell Laboratories.

De Sarbo, W.S., Manrai, A.K. and Burke, R.R. 1990: A nonspatial methodology for the analysis of two-way proximity data incorporating the distance-density hypothesis. *Psychometrika,* **55,** 229–53.

De Soete, G. 1984a: Additive tree representations of incomplete dissimilarity data. *Quality and Quantity,* **18,** 387–93.

De Soete, G. 1984b: A least squares algorithm for fitting an ultrametric tree to a dissimilarity matrix. *Pattern Recognition Letters,* **2,** 133–37.

De Soete, G. 1984c: Ultrametric tree representations of incomplete dissimilarity data. *Journal of Classification,* **1,** 235–42.

De Soete, G. and Carroll, J.D. 1983: A maximum likelihood method for fitting the wandering vector model. *Psychometrika,* **48,** 553–66.

De Soete, G. and Carroll, J.D. 1996: Trees and other network models for representing proximity data. *Clustering and Classification.* Arabie, P., Hubert, L. J. and De Soete, G. (Eds), pp. 157–97. New Jersey: World Scientific Publications.

De Soete, G. and Vermeulen, J. 1993: Interactively displaying ultrametric and additive trees. *Information and Classification.* O. Lausen, B. Opitz and R. Klar, (Eds), pp. 374–83. Berlin: Springer Verlag.

De Soete, G., De Sarbo, W.S., Furnas, G.W. and Carroll, J.D. 1984: The estimation of ultrametric and path length trees from rectangular proximity data. *Psychometrika,* **49,** 289–310.

De Soete, G., Carroll, J.D. and De Sarbo, W.S. 1986: The wandering ideal point model: a probabilistic multidimensional unfolding model for paired comparison data. *Journal of Mathematical Psychology,* **30,** 28–41.

Degerman, R.L. 1970: Multidimensional analysis of complex structure mixtures of class and quantitative variation. *Psychometrika,* **35,** 475–91.

Delbeke, L. 1968: *Construction of Preference Spaces.* Louvain: Publications of the University of Louvain.

Dobson, A.J. 1974: Unrooted trees for numerical taxonomy. *Journal of Applied Probability,* **11,** 32–42.

Eckes, T. and Orlik, P. 1991: An agglomerative method for two-mode hierarchical clustering. *Classification Data Analysis and Knowledge Organization,* pp. 7–8. Berlin: Springer Verlag.

Eisler, H. and Lindman, R. 1990: Representations of dimensional models of similarity. *Psychophysical Explanations of Mental Structures.* Geissler, H.O. (Ed), pp. 165–71. Toronto: Hogrefe and Huber.

Ekman, G. 1954: Dimensions of colour vision. *Journal of Psychology,* **38,** 467–74.

Espejo, E. and Gaul, W. 1986: Two-mode hierarchical clustering as an instrument for marketing research. *Classification as a Tool for Research,* pp. 121–28. Amsterdam: North-Holland.

Estabrook, G.F. and Rodgers, D.J. 1966: A general method of taxonomic description for a computed similarity measure. *Bioscience,* **16,** 789–93.

Everitt, B.S. 1993: *Cluster Analysis*. London: Arnold.

Everitt, B.S. and Gower, J.C. 1981: Plotting the optimal positions of an array of cortical electrical phosphenes. *Interpreting Multivariate data*. Chichester: Wiley.

Everitt, B.S. and Rushton, D.N. 1978: A method for plotting the optimum positions of an array of cortical electrical phosphenes. *Biometrics*, **34**, 399–410.

Fleiss, J.L. and Zubin, J. 1969: On the methods and theory of clustering. *Multivariate Behavioural Research*, **4**, 235–50.

Furnas, G.W. 1980: Objects and their features. The metric representation of two class data. PhD Thesis, Stanford University.

Genstat 5. 1993: *Genstat 5 Release 3 Reference Manual*. Oxford.

Gnanadesikan, R. and Wilk, M.B. 1969: Data analytic methods in multivariate statistical analysis. *Multivariate Analysis*, vol. 2. P.R. Krishnaiah, (Ed.), New York: Academic Press.

Goldenberger, G., Perdreka, I. Suess, E. and Descke, L. 1989: The cerebral localization of neuropsychological impairment in Alzheimer's disease: a SPECT study. *Journal of Neurology*, **236**, 131–38.

Goodall, C.R. 1991: Procrustes methods in the statistical analysis of shape. *Journal of the Royal Statistical Society, Series B*, **53**, 285–339.

Gordon, A.D. 1981: *Classification*. London: Chapman & Hall.

Gordon, A.D. 1987: A review of hierarchical classification. *Journal of the Royal Statistical Society, Series A*, **150**, 119–37.

Gordon, A.D. 1996: Hierarchical classification. *Clustering and Classification*. P. Arabie, L.J. Hubert and G. De Soete, (Eds), River-Edge, NJ: World Scientific Publications.

Gower, J.C. 1966a: A Q-technique for the calculation of canonical variates. *Biometrika*, **53**, 388–89.

Gower, J.C. 1966b: Some distance properties of latent root and vector methods used in multivariate analysis. *Biometrika*, **53**, 325–38.

Gower, J.C. 1971: Statistical methods for comparing different multivariate analyses of the same data. *Mathematics in the Archeological and Historical Sciences*. C.R. Modson, D.G. Kendall and P. Tautu, (Eds), pp. 138–49. Edinburgh: Edinburgh University Press.

Gower, J.C. 1975: Generalized Procrustes analysis. *Psychometrika*, **40**, 33–51.

Gower, J.C. 1977: The analysis of asymmetry and orthogonality. *Recent Developments in Statistics*. J.R. Barra, F. Brodeau, G. Romier and B. Van Cutsem, (Eds), Amsterdam: North-Holland.

Gower, J.C. and Hand, D.J. 1996: *Biplots*. London: Chapman and Hall.

Gower, J.C. and Legendre, P. 1986: Metric and Euclidean properties of dissimilarity coefficients. *Journal of Classification*, **5**, 5–48.

Gower, J.C. and Ross, G.J.S. 1969: Minimum spanning trees and single linkage cluster analysis. *Applied Statistics*, **18**, 54–64.

Greenacre, M.J. and Browne, M.W. 1973: An efficient least squares algorithm to perform multidimensional unfolding. *Psychometrika*, **51**, 241–50.

Groenen, P.J.F. and Heiser, W.J. 1996: The tunneling method for global optimization in multidimensional scaling. *Psychometrika*, **61**, 529–50.

Guttman, L. 1968: A general nonmetric technique for finding the smallest coordinate space for a configuration of points. *Psychometrika*, **33**, 469–506.

Harshman, R.A. 1972: Determination and proof of minimum uniqueness conditions for PARAFAC-1. *UCLA Working Papers in Phonetics*, **22**.

Harshman, R.A., Green, P.E., Wind, Y., and Lundy, M.E. 1982: A model for the analysis of asymmetric data in marketing research. *Marketing Science*, **1**, 205–42.

Hartigan, J.A. 1967: Representation of similarity matrices by trees. *Journal of the American Statistical Association*, **62**, 1140–58.

Hartigan, J.A. 1975: *Clustering Algorithms*. New York: John Wiley and Sons.

Henley, N.M. 1969: A psychological study of the semantics of animal terms. *Journal of Verbal Learning and Verbal Behaviour*, **8**, 176–84.

Horan, C.B. 1969: Multidimensional scaling: combining observations when individuals have different perceptual structures. *Psychometrika*, **34**, 139–65.

Hubert, L. and Arabie, P. 1995: The approximation of two-mode proximity matrices by sums of order-constrained matrices. *Psychometrika*, **60**, 573–605.

IMSL. 1987: *IMSL STAT/LIBRARY User's Manual*.

Jaccard, P. 1908: Nouvelles recherches sur la distribution florale. *Bulletin de la Societe Vaudoise de Sciences Naturelles*, **44**, 223–370.

Jacoby, W.G. 1991: *Data Theory and Dimensional Analysis*. Newbury Park, CA: Sage.

Jardine, N., and Sibson, R. 1971: *Mathematical Taxonomy*. London: Wiley.

Johnson, S.C. 1967: Hierarchical clustering schemes. *Psychometrika*, **32**, 241–54.

Kaufman, L. and Rousseeuw, P.J. 1990: *Finding Groups on Data*. New York: John Wiley and Sons.

Kendall, D.G. 1970: A mathematical approach to seriation. *Philosophical Transactions of the Royal Society of London*, **269**, 125–35.

Kendall, D.G. 1971: Seriation from abundance matrices. *Mathematics in the Archeological and Historical Sciences, pp. 215–52*. Edinburgh: Edinburgh University Press.

Kirk, G.S. 1974: *The Nature of Greek Myth*. Penguin Books.

Krause, E.F. 1975: *Taxicab Geometry*. Menlo Park, CA: Addison-Wesley.

Kruskal, J.B. 1964a: Multidimensional scaling by optimizing goodness of-fit to a nonmetric hypothesis. *Psychometrika*, **29**, 1–27.

Kruskal, J.B. 1964b: Nonmetric multidimensional scaling: a numerical method. *Psychometrika*, **29**, 28–42.

Kruskal, J.B. 1965: Analysis of factorial experiments by estimating monotone transformations of the data. *Journal of the Royal Statistical Society Series B*, **27**, 251–63.

Kruskal, J.B. and Carroll, J.D. 1969: Geometrical models and badness of fit functions. *International Multivariate Analysis, Dayton Ohio, pp. 639–70*. New York: Academic Press.

Kruskal, J.B. and Wish, M. 1978: *Multidimensional Scaling*. Beverly Hills, CA: Sage.

Kruskal, J.B., Young, F.W. and Seery, J.B. 1977: *How to Use KYST-2, a Very Flexible Program to do Multidimensional Scaling and Unfolding*. Unpublished manuscript, Bell Laboratories, Murray Hill, NJ.

Krzanowski, W.J. 1994: Ordination in the presence of group structure, for general multivariate data. *Journal of Classification*, **11**, 195–207.

Krzanowski, W.J. and Marriot, F.H.C. 1994: *Multivariate Analysis, Part I.* London: Arnold.

Krzanowski, W.J. and Marriot, F.H.C. 1995: *Multivariate Analysis, Part II.* London: Arnold.

Künnapas, T., Mälhammar, G. and Svenson, O. 1964: Multidimensional ratio scaling and multidimensional similarity of simple geometric figures. *Scandinavian Journal of Psychology*, 5, 249–56.

Kurczynski, T.W. 1969: *Genetic drift in a human isolate.* Ph.D. Thesis, Case Western University.

Kurczynski, T.W. 1970: Generalised distance and discrete variables. *Biometrics*, 26, 525–34.

Lance, G.N. and Williams, W.T. 1966: Computer programs for hierarchical polythetic classification. *Computer Journal*, 9, 60–4.

Langeheine, R. 1982: Statistical evaluation of measures of fit in the Lingoes-Borg Procrustean differences scaling. *Psychometrika*, 47, 427–42.

Larson, R.C., and Sadiq, G. 1983: Facility locations with the Manhattan metric in the presence of barriers to travel. *Operations Research*, 31, 652–99.

Lee, S.Y. 1984: Multidimensional scaling models with inequality and equality constraints. *Communications in Statistics, Part B. Simulation and Computation*, 13, 127–40.

Lee, S.Y. and Bentler, P.M. 1980: Functional relations in multidimensional scaling. *British Journal of Mathematical and Statistical Psychology*, 33, 142–50.

Legendre, L. and Legendre, P. 1983: *Numerical Ecology.* Amsterdam: Elsevier.

Legendre, P. and Chodorowski, A. 1977: A generalisation of Jaccard's association coefficient for Q-analysis of multi-state ecological data matrices. *Ekologia Poshha*, 25, 297–308.

Lingoes, J.C. 1966: An IBM-7090 program for Guttman–Lingoes smallest space analysis-RI. *Behavioral Science*, 11, 332.

Lingoes, J.C. 1971: Some boundary conditions for a monotone analysis of symmetric matrices. *Psychometrika*, 35, 195–203.

Lingoes, J.C. 1973: *The Guttman–Lingoes Nonmetric Program Series.* Ann Arbor: Mathesis Press.

Lingoes, J.C. and Borg, I. 1977: Optimale Lösungen für Dimensions- und Vektorgewichte in PINDIS. *Zeitschrift für Sozialpsychologie*, 8, 210–17.

Lingoes, J.C. and Borg, I. 1978: A direct approach to individual differences scaling using increasingly complex transformations. *Psychometrika*, 43, 491–519.

Lingoes, J.C. and Roskam, E.E. 1973: A mathematical and empirical analysis of multidimensional scaling algorithms. *Psychometrika Monograph Supplement*, 38, 1–93.

Loberman, H. and Weinberg, A. 1957: Formal procedures for connecting terminals with a minimum total path length. *Journal of the Association of Computer Machinery*, 4, 423–37.

MacCallum, R.C. 1977: Effects of conditionality of INDSCAL and ALSCAL weights. *Psychometrika*, 42, 297–305.

Mahalanobis, P.C. 1936: *Proceedings of the National Institute of Science of India*, 12, 49–55.

Manly, B.F.J. 1986: *Multivariate Statistical Methods*. London: Chapman and Hall.

Mardia, K.V. 1978: Some properties of classical multi-dimensional scaling. *Communications in Statistics. Theory and Methods*, **A7(13)**, 1233–41.

Mardia, K.V., Kent, J.T. and Bibby, J.M. 1979: *Multivariate Analysis*. London: Academic Press.

Marks, W.B. 1965: Difference spectra of the visual pigmentation in single goldfish cones. PhD Thesis, Johns Hopkins University.

McGee, V.C. 1968: Multidimensional scaling of n sets of similarity measures: a nonmetric individual difference approach. *Multivariate Behavioral Research*, **3**, 233–48.

Meulman, J.J. 1986: *A Distance Approach to Nonlinear Multivariate Analysis*. London: DSWO Press.

Meulman, J.J. 1992: The integration of multidimensional scaling and multivariate analysis with optimal transformations. *Psychometrika*, **57**, 539–65.

NAG. 1993: *The NAG Fortran Library Manual, Mark 16*. Oxford.

Nathanson, J.A. 1971: Applications of multivariate analysis in astronomy. *Applied Statistics*, **20**, 239–49.

Patrinos, A.N. and Hakimi, S.L. 1972: The distance matrix of a graph and its tree realization. *Quarterly Journal of Applied Mathematics*, **30**, 255–69.

Pollock, V.E. and Cliff, N. 1992: Analysis of pattern reversal and visual evoked potential topograhy reliability in normal subjects and comparisons with Alzheimer's disease. *Psychophysica*, **29**, 712–33.

Powell, M.J.D. 1977: Restart procedures for the conjugate gradient method. *Mathematical Programming*, **12**, 241–54.

Prim, R.C. 1957: Shortest connection matrix network and some generalizations. *Bell System Technical Journal*, **36**, 1389–1401.

Pruzansky, S. 1975: *How to use SINDSCAL, a computer program for individual differences in multidimensional scaling*. Unpublished manuscript, Murray Hill, NJ: Bell Laboratories.

Pruzansky, S., Tversky, A. and Carroll, J.D. 1982: Spatial versus tree representations of proximity data. *Psychometrika*, **47**, 3–24.

Rabinowitz, G.B. 1976: A procedure for ordering object pairs consistent with the multidimensional unfolding model. *Psychometrika*, **41**, 349–473.

Ramsay, J.O. 1977: Maximum likelihood estimation in multidimensional scaling. *Psychometrika*, **42**, 241–66.

Ramsay, J.O. 1978a: Confidence regions for multidimensional scaling analysis. *Psychometrika*, **43**, 145–60.

Ramsay, J.O. 1978b: *MULTISCALE: Four Programs of Multidimensional Scaling by Maximum Likelihood*. Chicago: International Educational Services.

Ramsay, J.O. 1980: Some small sample results for maximum likelihood estimation in maximum likelihood multidimensional scaling. *Psychometrika*, **45**, 139–44.

Ramsay, J.O. 1982: Some statistical approaches to multidimensional scaling data. *Journal of the Royal Statistical Society, Series A*, **145**, 285–312.

Ramsay, J.O. 1991: *Multiscale Manual (Extended Version)*. Canada: McGill University.

Rao, C.R. 1948: Tests of significance in multivariate analysis. *Biometrika*, **35**, 58–79.

Rao, C.R. 1949: On the distance between two populations. *Sankya*, **9**, 246–48.

Rodgers, J.L. and Young, F.W. 1981: Successive unfolding of family preferences. *Applied Psychological Measurement*, **5**, 51–62.

Rogers, D.J. and Tanimoto, T.T. 1960: A computer program for classifying plants. *Science*, **132**, 1115–1118.

Rosenberg, S. 1982: The method of sorting in multivariate research with applications selected from cognitive psychology and person perception. *Multivariate Applications in the Social Sciences*. N. Hirschberg and L.G. Humphreys, (Eds), pp. 117–42. New York: Erlbaum.

Rosenberg, S. and Kim, M.P. 1975: The method of sorting as a data-gathering procedure in multivariate research. *Multivariate Behavioural Research*, **10**, 489–502.

Roskam, E.E. and Lingoes, J.C. 1970: MINISSA-I: A Fortran IV (G) program for smallest space analysis of square symmetric matrices. *Behavioral Science*, **15**, 204–05.

Ross, J. 1966: A remark on Tucker and Messik's 'points of view' analysis. *Psychometrika*, **31**, 27–31.

Rothkopf, E.Z. 1957: A measurement of stimulus similarity and errors in some paired-associate learning tasks. *Journal of Experimental Psychology*, **53**, 94–101.

Sammon, J.W. 1969: A nonlinear mapping for data structure anaysis. *IEEE Transactions on Computers*, **18**, 401–09.

SAS 1966: *SAS/STAT Software Changes and Enhancements through Release 6.11*. Cary, NC.

Sattath, S. and Tversky, A. 1977: Additive similarity trees. *Psychometrika*, **42**, 319–45.

Schiffman, S.S., Reynolds, M.L. and Young, F.W. 1981: *Introduction to Multidimensional Scaling: Theory, Methods, and Applications*. New York: Academic Press.

Schönemann, P.H. 1970: On metric multidimensional unfolding. *Psychometrika*, **35**, 349–66.

Shepard, R.N. 1962a: The analysis of proximities: multidimensional scaling with unknown distance function Part I. *Psychometrika*, **27**, 125–40.

Shepard, R.N. 1962b: The analysis of proximities: multidimensional scaling with unknown distance function Part II. *Psychometrika*, **27**, 219–46.

Shepard, R.N. 1974: Representation of structure in similarity data problems and prospects. *Psychometrika*, **39**, 373–421.

Shepard, R.N. 1980: Multidimensional scaling, tree-fitting and clustering. *Science*, **210**, 340–98.

Sibson, R. 1979: Studies in the robustness of multidimensional scaling. Perturbational analysis of classical scaling. *Journal of the Royal Statistical Society, Series B*, **41**, 217–29.

Sokal, R.R. 1961: Distance as a measure of taxonomic similarity. *Systematic Zoology*, **10**, 70–9.

Sokal, R.R. and Sneath, P.H. 1963: *Principles of Numerical Taxonomy*. London: Freeman and Co.

SOLO 1991: *SOLO Categorical Data Analysis.*

Spence, I. 1970: *Multidimensional Scaling; an Empirical and Theoretical Investigation.* PhD Thesis, University of Toronto.

Spence, I. 1972: An aid to the estimation of dimensionality in nonmetric multidimensional scaling. *University of Western Ontario Research Bulletin,* 229.

Spence, I. and Graef, J. 1974: The determination of the underlying dimensionality of an empirically obtained matrix of proximities. *Multivariate Behavioural Research,* **9**, 311–42.

Spence, I. and Lewandowsky, S. 1989: Robust multidimensional scaling *Psychometrika,* **54**, 501–13.

SPSS 1994: *SPSS Professional Statistics 6.1.* Chicago, Illinois.

Storms, G. 1995: On the robustness of maximum likelihood scaling for violations of the error model. *Psychometrika,* **60**, 247–58.

SUGI 1986: *SUGI Supplemental Library User's Guide, Version 5 ed.* Cary, NC.

SYSTAT 1996: *SYSTAT 6.0 for Windows; Statistics.* Chicago, IL.

Takane, Y. and Sergent, J. 1983: Multidimensional scaling models for reaction times and same-different judgements. *Psychometrika,* **48**, 393–423.

Takane, Y., Young, F.W. and De Leeuw, J. 1977: Nonmetric individual differences multidimensional scaling: an alternating least squares method with optimal scaling features. *Psychometrika,* **42**, 7–67.

Torgerson, W.S. 1952: Multidimensional scaling 1: theory and method. *Psychometrika,* **17**, 401–19.

Torgerson, W.S. 1958: *Theory and Methods of Scaling.* New York: Wiley.

Tucker, L.R. 1960: Intra-individual and inter-individual multidimensionality. *Psychological Scaling: Theory and Applications,* H. Gulliksen and S. Messick, (Eds), pp. 109–27. New York: Wiley.

Tucker, L.R. 1972: Relations between multidimensional scaling and three mode factor analysis. *Psychometrika,* **37**, 3–27.

Tucker, L.R. and Messick, S. 1963: An individual differences model for multidimensional scaling. *Psychometrika,* **28**, 333–67.

Wagenaar, W.A. and Padmos, P. 1971: Quantitative interpretation of stages in Kruskal's multidimensional scaling technique. *British Journal of Mathematical and Statistical Psychology,* **24**, 101–10.

Weeks, D.G., and Bentler, P.M. 1982: Restricted multidimensional scaling models for asymmetric proximities. *Psychometrika,* **39**, 201–08.

Weinberg, S.L. and Menil, V.C. 1993: The recovery of structure in linear and ordinal data. *Multivariate Behavioural Research,* **28**, 215–33.

Young, F.W. 1972: A model for conjoint polynomial analysis algorithms. *Multidimensional Scaling, vol. 1,* R.N. Shepard, A.K. Romney and S.B. Nerlove, (Eds), pp. 105–53. New York: Seminar Press.

Young, F.W. 1974: Scaling replicated conditional rank-order data. *Sociological Methodology,* D. Heise, (Ed), American Sociological Association.

Young, F.W. 1975: An asymmetric Euclidean model for multi-process asymmetric data. *US–Japan Seminar on Theory, Methods and Applications of Multidimensional Scaling and Related Techniques.* University of California, San Diego, La Jolla.

Young, F.W. 1982: Enhancements in ALSCAL-82. *SUGI Proceedings,* **7**, 633–42.

Young, F.W. 1987: *Multidimensional Scaling; History, Theory, and Applications*. Hillsdale, New Jersey; Lawrence Erlbaum Associates, Inc.

Young, F.W. 1994: Multidimensional scaling. *SPSS Professional Statistics 6.1.* M.J. Norušis, (Ed), Chicago, Illinois: SPSS Inc.

Young, F.W. and Lewyckyi, R. 1979: *ALSCAL 4 Users' Guide*. Carrboro, NC: Data Analysis and Theory Associates.

Young, F.W. and Torgerson, W.S. 1967: TORSCA: A FORTRAN IV program for Shepard–Kruskal multidimensional scaling. *Behavioral Science*, **12**, 498.

Young, G. and Householder, A.S. 1938: Discussion of a set of points in terms of their mutual distances. *Psychometrika*, **3**, 19–22.

Zielman, B. and Heiser, W.J. 1996: Models for asymmetric proximities. *British Journal of Mathematical and Statistical Psychology*, **49**, 127–46.

Zinnes, J.L. and Griggs, R.A. 1974: Probabilistic multidimensional unfolding analysis. *Psychometrika*, **39**, 327–350.

Zinnes, J.L. and MacKay, R.A. 1983: Probabilistic multidimensional scaling: complete and incomplete data. *Psychometrika*, **48**, 27–48.

Author Index

Subject Index